Lecture Notes in Computer Science   14485

The series Lecture Notes in Computer Science (LNCS), including its subseries Lecture Notes in Artificial Intelligence (LNAI) and Lecture Notes in Bioinformatics (LNBI), has established itself as a medium for the publication of new developments in computer science and information technology research, teaching, and education.

LNCS enjoys close cooperation with the computer science R & D community, the series counts many renowned academics among its volume editors and paper authors, and collaborates with prestigious societies. Its mission is to serve this international community by providing an invaluable service, mainly focused on the publication of conference and workshop proceedings and postproceedings. LNCS commenced publication in 1973.

Javier Cámara · Sung-Shik Jongmans
Editors

Formal Aspects of Component Software

19th International Conference, FACS 2023
Virtual Event, October 19–20, 2023
Revised Selected Papers

 Springer

Editors
Javier Cámara (ID)
University of Malaga
Málaga, Spain

Sung-Shik Jongmans (ID)
Open University of Netherlands
Heerlen, The Netherlands

ISSN 0302-9743   ISSN 1611-3349 (electronic)
Lecture Notes in Computer Science
ISBN 978-3-031-52182-9   ISBN 978-3-031-52183-6 (eBook)
https://doi.org/10.1007/978-3-031-52183-6

This Springer imprint is published by the registered company Springer Nature Switzerland AG
The registered company address is: Gewerbestrasse 11, 6330 Cham, Switzerland

Paper in this product is recyclable.

Message from the PC Chairs

This volume contains the papers presented at the 19th International Conference on Formal Aspects of Component Software (FACS 2023), held online during October 19–20, 2023.

FACS aims to bring together practitioners and researchers in the areas of component software and formal methods in order to promote a deeper understanding of how formal methods can or should be used to make component-based software development succeed. The component-based software development approach has emerged as a promising paradigm to transport sound production and engineering principles into software engineering and to cope with the ever-increasing complexity of present-day software solutions. However, many conceptual and technological issues remain in component-based software development theory and practice that pose challenging research questions. Moreover, the advent of digitalization and industry 4.0, which requires better support from component-based solutions, e.g., cloud computing, cyber-physical and critical systems, and the Internet of Things, has brought to the fore new dimensions. These include quality of service, safety, and robustness to withstand inevitable faults, which require established concepts to be revisited and new ones to be developed in order to meet the opportunities offered by these supporting technologies.

To celebrate the 20th anniversary of FACS, this edition invited submissions to a special track on the topic of "component-based systems through the years" that describe important results and success stories that originated in the context of component-based software engineering.

The research track received 18 submissions, out of which the Program Committee selected 6 papers (33% acceptance rate), whereas all five invited submissions to the anniversary track were accepted. All submitted papers were reviewed by three referees (single blind). These proceedings include all the final versions of the accepted papers, taking into account the comments received by the reviewers. Authors of selected accepted papers will be invited to submit extended versions of their contributions to appear in a special issue of Elsevier's journal Science of Computer Programming.

We would like to thank all researchers who submitted their work to the conference, the Steering Committee members who provided precious guidance and support, all colleagues who served on the Program Committee, and the external reviewers, who helped us to prepare a high-quality conference program. Particular thanks to the invited speakers, Marsha Chechik from the University of Toronto in Canada and Rajeev Alur from the University of Pennsylvania in the USA, for their efforts and dedication to present their research and to share their perspectives on formal methods at FACS. We are extremely grateful for their help in managing practical arrangements to the Open University of the Netherlands, to Springer for their sponsorship, and to FCT, the Portuguese Foundation for Science and Technology, within the project IBEX, with reference PTDC/CCI-COM/4280/2021, for additional financial support.

November 2023   Javier Cámara
                                                                 Sung-Shik Jongmans

Message from the Anniversary Chair: Exploring the Available FACS Impact (2003–2023)

Joint work with Louis Robert

Introduction. This preface is dedicated to a bibliometric analysis of the proceedings of the first eighteen editions of the International Conference on Formal Aspects of Component Software (FACS), also known as "workshop" and "symposium" for its first ten editions, to perceive its achievements and evaluate its impact on the community.

Methodology. Bibliographic and bibliometric data were collected from different but complementary resources. Three databases[1]–Dimensions, Scopus and Web of Science– allowed a first visualization of data, and afterwards complete extractions of the elements necessary for the study. In addition, to extend the period covered by those databases, search engines[2]–such as Lens, Semantic Scholar and openAIRE–were used. Moreover, these resources have been complemented by consultation using SpringerLink, ScienceDirect and DBLP[3] for monitoring and enumeration purposes, namely by using digital object identifiers, DOI.

Our search queries, although adapted to the resource and engine used, were in general centred on the full title of the conference and its acronym (FACS); most interfaces require the use of the all search criterion (search in all fields), and of quotes to constrain the expression. The changing status of meetings (workshop, symposium and then conference) was covered by a regular expression when it was possible. In addition, let us note that most of the platforms require information concerning the document type, a disciplinary filter (e.g., computer science) and a filter on the period (2003–2023).

Results. The FACS proceedings, either *Electronic Notes in Theoretical Computer Science* (ENTCS, 2005–2009) or *Lecture Notes in Computer Science* (LNCS, 2010–2022), contain 294 contributions by 476 authors from 39 countries. Starting from 2005, each volume contains 10 (2021) to 23 (2005, 2013, 2014) articles. In addition, 75 selected extended papers have been published in 13 special issues of *Science of Computer Programming* (SCP, 2010-2022) and *Software and System Modeling* (SoSyM, 2023) international journals. Let us note that in the databases used, the ENTCS volumes are not

---

[1] https://www.dimensions.ai.
  https://www.scopus.com.
  https://www.webofscience.com.
[2] https://www.lens.org.
  https://www.semanticscholar.org.
  https://explore.openaire.eu.
[3] https://link.springer.com.
  https://www.sciencedirect.com.
  https://dblp.org.

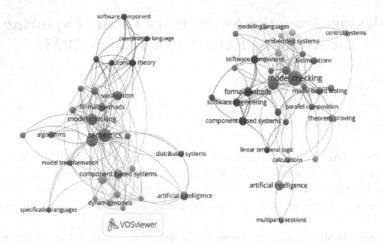

Fig. 1. Keywords co-occurrency: 2011–2016 vs. 2017–2022

explicitly linked to the FACS workshop, making extractions partial with 170 records at most. On their side, the (manually built) DOI-based queries provide 263 records.

The data that caught our attention are the titles, keywords and abstracts. They have been the subject of text mining to highlight the topics and the themes covered, and their evolution over the two FACS decades. The data processing was carried out thanks to the available export and analysis functions for the databases and the engines used. For result visualization, the VOSViewer tool[4] was also used, like in Fig. 1, where VOS stands for visualization of similarities.

By comparing the co-occurrence of the keywords with VOSviewer using a technique that categorizes the keywords in clusters, we found some interesting features of the topics landscape over the past two decades. First, formal aspects of component-based system development–design, specification, verification and validation–stay central to the FACS community over the years. Second, whereas the focus of the first editions was mainly on software components, it was then put on distributed, embedded and complex systems in general, and more recently on cyber-physical systems. Third, the artificial intelligence-related topics were diffuse in 2015–2016, but now the artificial intelligence cluster is formed and anchored to systems' analysis activities.

Discussion. Bibliometric data provide a perspective on the visibility of the FACS conference. Obviously, it depends on databases' core collections. Open Access (OA) also provides a perspective for analysis on how important it has been over the last twenty years, and on how OA has increased the visibility of the FACS contributions. The data on OA available in the databases, which are displayed in Fig. 2, together with the possibility of sorting by citation, allowed us to provide a first attempt to address these questions, with the creation of top lists of the most-cited FACS articles (top 10, top 20 and top 30). For all the databases explored, more than 50% of the FACS contributions are in OA. Let

---

[4] N. J. van Eck and L. Waltman. Software survey: VOSviewer, a computer program for bibliometric mapping. *Scientometrics*, 84(2):523–538, 2010.

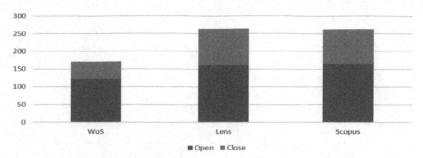

Fig. 2. Open Access: Synthesis for WoS, Lens, and Scopus databases

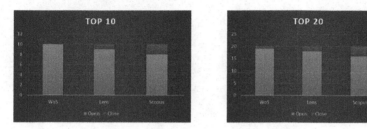

Fig. 3. Open Access for 10 and 20 most cited articles

us emphasize that the OA ratio for the 3 lists of the most-cited articles is greater then for all the articles in general, as illustrated in Fig. 3 for the top 10 and top 20 articles in comparison with Fig. 2.

Currently, the overall impact analysis is mainly based on citations for both databases and search engines. A more precise impact analysis for separate contributions is possible, e.g., with Semantic Scholar to identify influential citations. Notice that it also depends on the access to the full text of the citing papers in the core collections.

November 2023   Olga Kouchnarenko

Organization

Program Committee

Antónia Lopes	Universidade de Lisboa, Portugal
Anton Wijs	Eindhoven University of Technology, The Netherlands
Arpit Sharma	Indian Institute of Science Education and Research, India
Brijesh Dongol	University of Surrey, UK
Camilo Rocha	Pontificia Universidad Javeriana Cali, Colombia
Clemens Dubslaff	Eindhoven University of Technology, The Netherlands
Emilio Tuosto	Gran Sasso Science Institute, Italy
Farhad Arbab	CWI/Leiden University, The Netherlands
Fatemeh Ghassemi	University of Tehran, Iran
Genaina Rodrigues	University of Brasilia, Brazil
Giorgio Audrito	University of Turin, Italy
Gwen Salaün	Grenoble Alpes University, France
Huibiao Zhu	East China Normal University, China
Ivan Lanese	University of Bologna, Italy
Jacopo Mauro	University of Southern Denmark, Denmark
Javier Cámara	University of Málaga, Spain/University of York, UK
José Proença	CISTER-ISEP/HASLab-INESC TEC, Portugal
Keigo Imai	Gifu University, Japan
Kenneth Johnson	Auckland University of Technology, New Zealand
Kyungmin Bae	Pohang University of Science and Technology (POSTECH), South Korea
Ludovic Henrio	CNRS, France
Luís Soares Barbosa	University of Minho, Portugal
Marie Farrell	University of Manchester, UK
Meng Sun	Peking University, China
Mieke Massink	CNR-ISTI, Pisa, Italy
Olga Kouchnarenko	University of Franche-Comté, France
Peter Ölveczky	University of Oslo, Norway
Rob van Glabbeek	University of Edinburgh, UK
Samir Genaim	Universidad Complutense de Madrid, Spain
Shoji Yuen	Nagoya University, Japan

Sung-Shik Jongmans   Open University of the Netherlands/CWI,
                                      The Netherlands
Violet Ka I Pun   Western Norway University of Applied Sciences,
                                      Norway
Zhiming Liu   Southwest University, China

Program Chairs

Javier Cámara   University of Málaga, Spain/University of York,
                                      UK
Sung-Shik Jongmans   Open University/CWI, The Netherlands

FACS 20th Anniversary Chair

Olga Kouchnarenko   University of Franche-Comté, France

Steering Committee

Farhad Arbab   CWI/Leiden University, The Netherlands
Kyungmin Bae   Pohang University of Science and Technology,
                                      South Korea
Peter Csaba Ölveczky   University of Oslo, Norway
Javier Cámara   University of Málaga, Spain/University of York,
                                      UK
Sung-Shik Jongmans   Open University of the Netherlands/CWI,
                                      The Netherlands
Zhiming Liu   Southwest University, China
Markus Lumpe   Swinburne University of Technology, Australia
Eric Madelaine   Inria Sophia Antipolis, France
Corina Pasareanu   CMU, USA
José Proença   Polytechnic Institute of Porto, Portugal
Gwen Salaün   Université Grenoble Alpes, France
Luís Soares Barbosa (Chair)   University of Minho, Portugal
Anton Wijs   Eindhoven University of Technology,
                                      The Netherlands

Additional Reviewers

Carwehl, Marc
Ciancia, Vincenzo
Delahaye, Benoit
Kaarsgaard, Robin
Li, Zhaokai
Mezzina, Claudio Antares
Osama, Muhammad
Sun, Weidi
Yang, Zhibin

Keynotes

Model Checking for Safe Autonomy

Rajeev Alur

University of Pennsylvania

We focus on the problem of formally verifying correctness requirements of a closed-loop control system where the controller is trained using machine learning. As an illustrative realistic case study, we consider an autonomous car that navigates a structured environment using a neural-network-based controller. In this scenario, safety corresponds to avoiding collisions, and we first discuss how to formalize this as a verification problem. Then we describe a specific solution strategy, advocated by the verification tool Verisig, which relies on tools for computing reachable states of hybrid dynamical systems. We conclude by discussing challenges and opportunities of applying formal verification to establish safety of autonomous systems with learning-enabled components.

Assurance for Software Product Lines Through Lifting and Reuse

Marsha Chechik

University of Toronto

From financial services platforms to social networks to vehicle control, complex software has come to mediate many activities of daily life. Software failures can have significant consequences to individuals, organizations and societies. As such, stakeholders require evidence-based assurance that software satisfies key requirements – for instance, that it is safe, secure, or protects privacy. Evidence is often generated using testing or verification techniques, making creation and maintenance of assurance an expensive process. Furthermore, many industries – from automotive to aerospace to consumer electronics – develop and maintain complex families of software systems in the form of product lines, which can yield billions of distinct products. It is infeasible to create evidence and assure each possible product individually; thus, there is a clear need for new approaches for assuring software product lines.

In this talk, I will discuss approaches for reuse of analyses and evidence through the formal process of lifting. I will provide an overview of recent results in this space and identify future challenges.

Contents

Research Papers

Symbolic Path-Guided Test Cases
for Models with Data and Time

Boutheina Bannour[1](✉), Arnault Lapitre[1], Pascale Le Gall[2],
and Thang Nguyen[2]

[1] Université Paris-Saclay, CEA, List, 91120 Palaiseau, France
boutheina.bannour@cea.fr
[2] Université Paris-Saclay, CentraleSupélec, MICS, 91192 Gif-sur-Yvette, France

Abstract. This paper focuses on generating test cases from timed symbolic transition systems. At the heart of the generation process are symbolic execution techniques on data and time. Test cases look like finite symbolic trees with verdicts on their leaves and are based on a user-specified finite symbolic path playing the role of a test purpose. Generated test cases handle data involved in time constraints and uninitialized parameters, leveraging the advantages of symbolic execution techniques.

Keywords: model-based testing · timed input/output symbolic transition systems · symbolic execution · tioco conformance relation · test purpose · test case generation · uninitialized parameters

1 Introduction

Context. Symbolic execution [7,13,15,20] explores programs or models' behaviors using formal parameters instead of concrete values and computes a logical constraint on them, the so-called path condition. Interpretations of these parameters satisfying the constraint yield inputs that trigger executions along the desired path. Symbolic execution's primary application is test case generation, where considering test cases guided by different symbolic paths facilitates achieving high coverage across diverse behaviors. Symbolic execution has been defined first for programs [20] and extended later to models [3,4,8,13,15,27] in particular to symbolic transition systems where formal parameters abstract values of uninitialized data variables [4,15], values of received data from the system's environment [1,3,4,8,13,15,27], and durations stored in clock variables [4,27].

Contribution. In this paper, we investigate test case generation from models given as symbolic transition systems that incorporate both data and time. Time is modeled with explicit clock variables, which are treated as a particular case of data variables that occur in guards and constrain the transitions' firing. Our approach allows for general logical reasoning that mixes data and time through symbolic execution, typically compared to Timed Automata [2], which are models dedicated to time and use tailored zone-based abstraction techniques to handle time. Test cases are built based on a test purpose, defined as a selected symbolic path of the model. We require test purposes to be deterministic, i.e. any system

J. Cámara and S.-S. Jongmans (Eds.): FACS 2023, LNCS 14485, pp. 3–22, 2024.
https://doi.org/10.1007/978-3-031-52183-6_1

behavior expressed as a trace cannot be executed both on the test purpose and on another symbolic path. By leveraging this determinism property and symbolic execution techniques, we define test cases as tree-like structures [19,24], presenting the following advantages: (i) data and time benefit from comparable property languages, seamlessly handled with the same symbolic execution techniques; (ii) input communication channels are partitioned into controllable input channels enabling the test case to stimulate the system under test, and uncontrollable input channels enabling observation of data from third parties; (iii) state variables do not need to be initialized, and finally (iv), test cases can be easily executed on systems under test, typically achieved through behavioral composition techniques, such as employing TTCN-3 [26], or by maintaining a test case state at runtime using on-the-fly test case execution [8,12,15]. In either case, our test cases are coupled with constraint solving to assess the satisfiability of test cases' progress or verdict conditions. We provide a soundness result of our test case execution on the system under test in the framework of the timed conformance relation tioco [21] issued from the well-established relation ioco [28]. Finally, we implement our test case generation in the symbolic execution platform Diversity [9].

Paper Plan. We devote Sect. 2 to present timed symbolic transition systems mixing data and time, and in Sect. 3, we define their symbolic execution serving as the foundation for our test case generation. In Sect. 4, we give the main elements of the testing framework: the conformance relation tioco and test purposes. In Sect. 5, we detail the construction of symbolic path-guided test cases. In Sect. 6, we provide some links to related work. In Sect. 7, we provide concluding words.

2   Timed Input/Output Symbolic Transition Systems

Preliminaries on Data Types. For two sets A and B, we denote B^A, the set of applications from A to B. We denote $\coprod_{i \in \{1,...n\}} A_i$ the disjoint union of sets A_1, ..., and A_n. For a set A, A^* (resp. A^+) denotes the set of all (resp. non-empty) finite sequences of elements of A, with ε being the empty sequence. For any two sequences $w, w' \in A^*$, we denote $w.w' \in A^*$ their concatenation.

A data signature is a pair $\Omega = (S, Op)$ where S is a set of type names and Op is a set of operation names provided with a profile in S^+. We denote $V = \coprod_{s \in S} V_s$ the set of typed variables in S with $type : V \to S$ the function that associates variables with their type. The set $T_\Omega(V) = \coprod_{s \in S} T_\Omega(V)_s$ of Ω-terms in V is inductively defined over V and operations Op of Ω as usual and the function $type$ is extended to $T_\Omega(V)$ as usual. The set $\mathcal{F}_\Omega(V)$ of typed equational Ω-formulas over V is inductively defined over the classical equality and inequality predicates $t \bowtie t'$ with $\bowtie \in \{<, \leq, =, \geq, >\}$ for any $t, t' \in T_\Omega(V)_s$ and over usual Boolean constants and connectives $True$, $False$, $\neg$, $\vee$, $\wedge$ and quantifiers $\forall x, \exists x$ with x a variable of V. We may use the syntax $\exists\{x_1, \ldots, x_n\}$ for the expression $\exists x_1 \ldots \exists x_n$. A substitution over V is a type-preserving application $\rho : V \to T_\Omega(V)$. The identity substitution over V is denoted id_V and substitutions are canonically extended on terms and formulas.

An Ω-model $M = (\coprod_{s \in S} M_s, (f_M)_{f \in Op})$ provides a set of values M_s for each type s in S and a concrete operation $f_M : M_{s_1} \times \cdots \times M_{s_n} \to M_s$ for each operation name $f : s_1 \ldots s_n \to s$ in Op. An interpretation $\nu : V \to M$ associates a value in M with each variable $v \in V$, and is canonically extended to $\mathcal{T}_\Omega(V)$ and $\mathcal{F}_\Omega(V)$ as usual. For ν an interpretation in M^V, x a variable in V and v a value in M, $\nu[x \mapsto v]$ is the interpretation $\nu' \in M^V$ which sends x on the value v and coincides with ν for all other variables in V. For $\nu \in M^V$ and $\varphi \in \mathcal{F}_\Omega(V)$, the satisfaction of φ by ν is denoted $M \models_\nu \varphi$ and is inductively defined w.r.t. the structure of φ as usual. We say a formula $\varphi \in \mathcal{F}_\Omega(V)$ is satisfiable, denoted $Sat(\varphi)$, if there exists $\nu \in M^V$ such that $M \models_\nu \varphi$.

In the sequel, we consider a data signature $\Omega = (S, Op)$ with $time \in S$ to represent durations and Op containing the usual operations $< : time.time \to Bool$ and $+ : time.time \to time, \ldots$ An Ω-model M being given, M_{time} is denoted D and is isomorphic to the set of non-negative real numbers. $< : time.time \to Bool$ and $+ : time.time \to time$ are mapped to their usual meanings.

Timed Input/Output Symbolic Transition Systems (TIOSTS) are automata handling data and time, and defined over a *signature* $\Sigma = (A, K, C)$, where:

- $A = \coprod_{s \in S} A_s$ and K are pairwise disjoint sets of variables representing respectively *data variables* and *clock variables* of type *time*
- and $C = \coprod_{s \in S} C_s$ is a set of *communication channels* with the convention $type(c) = s$ for any $c \in C_s$. Moreover, channels of type $s \in S$ are partitioned into input and output channels, i.e., $C_s = C_s^{in} \coprod C_s^{out}$.

We denote $C^{in} = \coprod_{s \in S} C_s^{in}$, resp. $C^{out} = \coprod_{s \in S} C_s^{out}$, the set of all input, resp. output, channels, regardless of their type.

Interactions of TIOSTS with the environment are expressed in terms of communication actions. The *set of communication actions* over Σ is $Act(\Sigma) = I(\Sigma) \cup O(\Sigma)$ where:

- $I(\Sigma) = \{c?x \mid c \in C^{in}, x \in A_{type(c)}\}$ is the set of input actions, and
- $O(\Sigma) = \{c!t \mid c \in C^{out}, t \in \mathcal{T}_\Omega(A \cup K)_{type(c)}\}$ is the set of output actions.

$c?x$ denotes the reception of a value to be stored in x through channel c. $c!t$ denotes the emission of the value corresponding to the current interpretation of term t through channel c. The *set of concrete communication actions* over C is $Act(C) = I(C) \cup O(C)$ where:

$I(C) = \{c?v \mid c \in C^{in}, v \in M_{type(c)}\}$ and $O(C) = \{c!v \mid c \in C^{out}, v \in M_{type(c)}\}$

Notations. For $a \in Act(\Sigma)$ (resp. $a \in Act(C)$) of the form $c?y$ or $c!y$, $chan(a)$ and $val(a)$ denote c and y respectively. For expressiveness concerns, we also use extensions of those actions: either carrying n pieces of data, i.e. $c!(t_1, \ldots, t_n)$ or $c?(x_1, \ldots, x_n)$, and simple signals $c!$ or $c?$ which are actions carrying no-data.

Definition 1 (TIOSTS). *A TIOSTS over $\Sigma = (A, K, C)$ is a triple $\mathbb{G} = (Q, q_0, Tr)$, where*

- *Q is the set of states,*
- *$q_0 \subset Q$ is the initial state,*

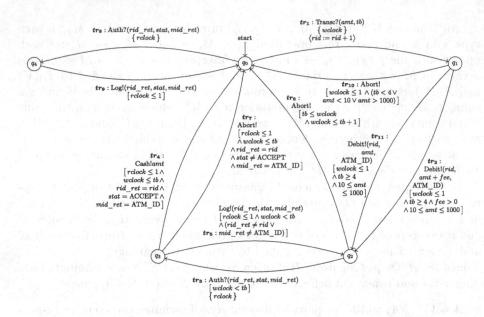

Fig. 1. Example TIOSTS of an ATM

- *Tr is the set of transitions of the form $(q, act, \phi, \mathbb{K}, \rho, q')$ with $q, q' \in Q$, $act \in Act(\Sigma)$, $\phi \in \mathcal{F}_\Omega(A \cup K)$, $\mathbb{K} \subseteq K$ and $\rho : A \to \mathcal{T}_\Omega(A \cup K)$ is a type-preserving function.*

In the sequel, given a transition tr of the form $(q, act, \phi, \mathbb{K}, \rho, q')$, we will access its components by their name, for example, $act(tr)$ for its communication action. We comment on the ingredients of a TIOSTS through the TIOSTS given in Example 1.

Example 1. The TIOSTS $\mathbb{G} = (Q, q_0, Tr)$ in Fig. 1 represents a simple Automatic Teller Machine (ATM) with $Q = \{q_0, \ldots, q_4\}$ and $Tr = \{tr_1, \ldots, tr_{11}\}$. Its signature introduces two clocks ($wclock$, $rclock$), 7 data variables (rid, amt, tb, fee, rid_ret, $stat$, mid_ret) and 6 channels including 2 input channels (Transc, Auth) and 4 output channels (Debit, Abort, Cash and Log).

Transition $tr_1 : q_0 \xrightarrow{\text{Transc?}(amt,tb),[True],\{wclock\},\langle rid:=rid+1\rangle} q_1$ represents a reception on channel Transc of a client withdrawal request for a given amount stored in variable amt and the corresponding bound on processing time stored in variable tb, which can vary due to bank security checks. tr_1 is unconditionally fired (due to the guard $True$), resets clocks in $\mathbb{K} = \{wclock\}$ and updates variable rid with $rid+1$. tr_1 abstracts client interaction and bank processing time retrieval. Transition tr_2 :
$$q_1 \xrightarrow{\text{Debit!}(rid,amt+fee,\text{ATM\_ID}),[wclock\leq 1\wedge tb\geq 4\wedge fee>0\wedge 10\leq amt\leq 1000],\{\},\langle\rangle} q_2$$
represents an emission of (bank) debit request on channel Debit of the value of rid, the value of the term $amt+fee$, and the value of the constant ATM_ID. tr_2

can be fired if and only if the duration since *wclock* reset is less than or equal 1, the processing time bound is greater than or equal 4, the ATM *fee* is strictly positive, and the withdrawal amount in some range (between 10 and 1000).

Other transitions represent debit authorization reception (tr_3), cash return (tr_4), logging non-involved debit authorization $(tr_5$ and $tr_9)$, cancellation upon timeout (tr_6) or debit refusal (tr_7), cancellation due to amount out of range or inappropriate processing time bound (tr_{10}), reception of non-involved debit authorization (tr_8), and feeless debit request (tr_{11}).

3   Symbolic Execution of TIOSTS

We use symbolic execution techniques for defining the semantics of TIOSTS: transitions are executed not for concrete values but rather using fresh variables and accumulating constraints on them. Given an TIOSTS $\mathbb{G} = (Q, q_0, Tr)$ over $\Sigma = (A, K, C)$, we consider a set F of fresh variables disjoint from TIOSTS variables, i.e. $F \cap (A \cup K) = \emptyset$, and partitioned with the following subsets:

- F^{ini} a set of variables dedicated to initialize variables of $\mathbb{G}$;
- $F^{in} = (F^{in}_c)_{c \in C^{in}}$ verifying that variables in F^{in}_c are of type $type(c)$;
- $F^{out} = (F^{out}_c)_{c \in C^{out}}$ verifying that variables in F^{out}_c are of type $type(c)$;
- F^{dur} a set of variables of type *time*.

For the signature $\Sigma_F = (F, \emptyset, C)$, the set $Evt(\Sigma_F)$ of *symbolic events* over Σ_F is $F_{time} \times (Act(\Sigma_F) \cup \{\_\})$ with $\_$ for indicating the absence of an action. For $ev = (z, act)$ in $Evt(\Sigma_F)$, $delay(ev)$ and $act(ev)$ denote resp. z and act. Intuitively, z is the duration elapsed between the action preceding act and act.

An Execution Context (EC) *ec* is a data structure of the form $(q, \pi, \lambda, ev, pec)$ composed of pieces of information about symbolic execution:

- $q \in Q$, a state (control point) of the TIOSTS reached so far,
- $\pi \in \mathcal{F}_\Omega(F)$, a constraint on variables in F, the so-called *path condition*, to be satisfiable by the symbolic execution to reach *ec*,
- $\lambda : A \cup K \to T_\Omega(F)$, a substitution,
- $ev \in Evt(\Sigma_F)$, a symbolic event that has been executed to reach *ec*,
- *pec* a predecessor of *ec* useful to build a symbolic tree in which nodes are ECs and edges connect predecessor ECs to ECs themselves.

For any execution context *ec*, we note $q(ec), \pi(ec), \lambda(ec), ev(ec)$ and $pec(ec)$ to denote the corresponding elements in *ec*. In the same line, we also note $act(ec)$, $delay(ec)$ and $chan(ec)$ for resp. $act(ev(ec))$, $delay(ev(ec))$ and $chan(act(ev(ec)))$. For convenience, $Sat(\pi(ec))$ will be denoted $Sat(ec)$.

Initial ECs are of the form $ec_0 = (q_0, True, \lambda_0, \_, self)$ with: λ_0 associating to every variable of A a distinct fresh variable of F^{ini}, and to variables of K the constant 0; "$\_$" an identifier indicating the absence of an action to start the system; and *self* an identifier indicating that , the predecessor of an initial EC is the initial context itself. $\mathbb{EC}(\mathbb{G})$ denotes the set of all ECs of a TIOSTS $\mathbb{G}$.

For a non-initial execution context ec, we use the notation $pec(ec) \xrightarrow{ev(ec)} ec$ or $pec(ec) \xrightarrow{ev(ec)} ec \in \mathbb{EC}$ if ec and its predecessor are both in a subset $\mathbb{EC}$ of $\mathbb{EC}(\mathbb{G})$.

Transitions are executed symbolically from an EC. An example execution on the TIOSTS of Fig. 1 is provided before presenting the general definition.

Example 2. The symbolic execution of transition tr_2 (given in Example 1) from execution context ec_1 results in a successor context ec_2 as follows:
$$ec_1 \xrightarrow{(\mathbf{z_1},\text{Debit}!(\mathbf{y_{D_1}^1},\mathbf{y_{D_2}^1},\mathbf{y_{D_3}^1}))} ec_2.$$

Figure 2 provides a summary of both contexts. In the execution context ec_1, the variables *fee*, *rid*, *rid_ret*, *stat*, and *mid_ret* are evaluated with fresh initial parameters. The successor context ec_2 is computed by associating the clock *wclock* with a new duration $\mathbf{z_1}$ in F^{dur}, indicating the time elapsed since the previous event.

The emission event $(\mathbf{z_1}, \text{Debit}!(\mathbf{y_{D_1}^1}, \mathbf{y_{D_2}^1}, \mathbf{y_{D_3}^1}))$ corresponds to the outcome of the symbolic evaluation of the transition action $act(tr_2) = \text{Debit}!(rid, amt + fee, \text{ATM\_ID})$. The variables $\mathbf{y_{D_1}^1}$, $\mathbf{y_{D_2}^1}$ and $\mathbf{y_{D_3}^1}$ are respectively in $F_{\text{Debit},1}^{out}$, $F_{\text{Debit},2}^{out}$ and $F_{\text{Debit},3}^{out}$

The evaluation of the transition guard $\phi(tr_2) = wclock \leq 1 \wedge tb \geq 4 \wedge fee > 0 \wedge 10 \leq amt \leq 1000$ yields the formula $\mathbf{z_1} \leq 1 \wedge \mathbf{tb_1} \geq 4 \wedge \mathbf{fee_0} > 0 \wedge 10 \leq \mathbf{amt_1} \leq 1000$ which together with identification conditions $\mathbf{y_{D_1}^1} = \mathbf{rid_0} + 1$, $\mathbf{y_{D_2}^1} = \mathbf{amt_1} + \mathbf{fee_0}$ and $\mathbf{y_{D_3}^1} = \text{ATM\_ID}$ constitutes $\pi(ec_2)$, the path condition of ec_2. Identification conditions result from the transition action evaluation.

Definition 2 will make clear the computation of the EC's successors from TIOSTS transitions. While in Example 2, we have illustrated the symbolic execution of a unique transition (tr_2), we will define symbolic execution by simultaneously executing all the outgoing transitions from a given EC. Following this approach, we can introduce the same symbolic variables for all outgoing transitions as far as they have the same role. Typically, the same fresh duration is employed to represent the time passing in the computation of all the EC's successors.

Generically, given an execution context ec in $\mathbb{EC}(\mathbb{G})$, we will access the symbolic variables introduced by the executions from ec with the following notations: $f_c^{in}(ec)$ for $c \in C^{in}$, $f_c^{out}(ec)$ for $c \in C^{out}$, and $f^{dur}(ec)$. For convenience, all such fresh variables are available by default with every execution

$q(ec_1) : q_1$
$tr(ec_1) : tr_1$
$pec(ec_1) : ec_0$
$\pi(ec_1) : True$
$\lambda(ec_1) : rid \mapsto \mathbf{rid_0} + 1, amt \mapsto \mathbf{amt_1},$
$\qquad fee \mapsto \mathbf{fee_0}, rid\_ret \mapsto \mathbf{rid\_ret_0},$
$\qquad stat \mapsto \mathbf{stat_0},$
$\qquad mid\_ret \mapsto \mathbf{mid\_ret_0},$
$\qquad tb \mapsto \mathbf{tb_1}, wclock \mapsto 0, rclock \mapsto \mathbf{z_0}$
$ev(ec_1) : (\mathbf{z_0}, \text{Transc}?(\mathbf{amt_1}, \mathbf{tb_1}))$
$q(ec_2) : q_2$
$tr(ec_2) : tr_2$
$pec(ec_2) : ec_1$
$\pi(ec_2) : (\mathbf{z_1} \leq 1)$
$\qquad \wedge (\mathbf{tb_1} \geq 4)$
$\qquad \wedge (\mathbf{fee_0} > 0)$
$\qquad \wedge (10 \leq \mathbf{amt_1} \leq 1000)$
$\qquad \wedge (\mathbf{y_{D_1}^1} = \mathbf{rid_0} + 1)$
$\qquad \wedge (\mathbf{y_{D_2}^1} = \mathbf{amt_1} + \mathbf{fee_0})$
$\qquad \wedge (\mathbf{y_{D_3}^1} = \text{ATM\_ID})$
$\lambda(ec_2) : rid \mapsto \mathbf{rid_0} + 1, amt \mapsto \mathbf{amt_1},$
$\qquad fee \mapsto \mathbf{fee_0}, rid\_ret \mapsto \mathbf{rid\_ret_0},$
$\qquad stat \mapsto \mathbf{stat_0}, tb \mapsto \mathbf{tb_1},$
$\qquad mid\_ret \mapsto \mathbf{mid\_ret_0},$
$\qquad wclock \mapsto \mathbf{z_1}, rclock \mapsto \mathbf{z_0} + \mathbf{z_1}$
$ev(ec_2) : (\mathbf{z_1}, \text{Debit}!(\mathbf{y_{D_1}^1}, \mathbf{y_{D_2}^1}, \mathbf{y_{D_3}^1}))$

Fig. 2. Symbolic execution of a TIOSTS transition

context ec, even if there is no outgoing transition from $q(ec)$ carrying on a given channel c. For $\alpha \in \{in, out\}$, $f^\alpha(ec) = \{f_c^\alpha(ec) \mid c \in C^\alpha\}$. In Def 2, to make it easier to read, $f_c^{in}(ec)$, $f_c^{out}(ec)$ and $f^{dur}(ec)$ are respectively denoted as x_c, y_c and z.

Definition 2 (Symbolic Execution of a TIOSTS). *Let* $\mathbb{G} = (Q, q_0, Tr)$ *be a TIOSTS,* $ec = (q, \pi, \lambda, ev, pec)$ *be an execution context in* $\mathbb{EC}(\mathbb{G})$, *and* x_c *be a fresh variable in* F_c^{in} *for any* $c \in C^{in}$, y_c *be a fresh variable in* F_c^{out} *for any* $c \in C^{out}$ *and let* z *be a fresh variable in* F^{dur}.

The successors of ec *are all execution contexts* ec' *of the form* $(q', \pi', \lambda', ev', ec)$ *verifying that there exists a transition* $tr = (q, act, \phi, \mathbb{K}, \rho, q')$ *in* Tr *and constituents* λ', ev' *and* π' *of* ec' *are defined as follows:*

- *the substitution* $\lambda' : A \cup K \rightarrow T_\Omega(F)$:

$$\lambda'(w) = \begin{cases} \lambda_0'(\rho(w)) & \text{if } w \in A \\ 0 & \text{if } w \in \mathbb{K} \\ \lambda_0'(w) & \text{else } i.e. \text{ if } w \in K \setminus \mathbb{K} \end{cases} \tag{1}$$

where $\lambda_0' : A \cup K \rightarrow T_\Omega(F)$ *is the auxiliary function defined as:*

$$\lambda_0'(w) = \begin{cases} x_c & \text{if } act = c?w \\ \lambda(w) + z & \text{if } w \in K \\ \lambda(w) & \text{else} \end{cases} \tag{2}$$

- ev' *is* $(z, c?x_c)$ *if* $act = c?w$ *and is* $(z, c!y_c)$ *if* $act = c!t$ *for a given channel* c
- π' *is the formula* $\pi \wedge \lambda_0'(\phi)$ *if* $act = c?w$ *and is* $\pi \wedge \lambda_0'(\phi) \wedge (y_c = \lambda_0'(t))$ *if* $act = c!t$ *for a given channel* c.

The symbolic execution $SE(\mathbb{G})$ *of* $\mathbb{G}$ *is a couple* $(ec_0, \mathbb{EC})$ *where:* ec_0 *is an arbitrary initial EC, and* $\mathbb{EC}$ *is the smallest set of execution contexts containing* ec_0 *and all successors of its elements.*

Most of the time (e.g., if it is possible to run a cycle of $\mathbb{G}$ for an arbitrarily long time), $\mathbb{EC}$ is an infinite set. The computation of the successors of an execution context translates the standard execution of a transition from that context: λ_0' is an intermediate substitution that advances all clocks by the same fresh duration z to indicate time passing, assigns to a data variable w a fresh variable x_c if w is the variable of a reception ($c?w$) and leaves the other data variables unchanged. Then, λ' is defined, for the data variables, by applying λ_0' on the terms defined by the substitution ρ introduced by tr and, for the clock variables, by resetting the variables of $\mathbb{K}$ to zero and advancing the other clocks using λ_0'. The event action is either $c?x_c$ (case of a reception $c?w$) or $c!y_c$ (case of an emission $c!t$). The path condition π' is obtained by the accumulation of the condition π of the predecessor context ec and of the guard ϕ of the transition evaluated using λ_0'. Moreover, in case of an emission, π' keeps track of the identification condition matching y_c with the evaluation $\lambda_0'(t)$ of the emitted term t.

In the sequel, we will denote $tr(ec)$ the transition that allows building the execution context ec. By convention, $tr(ec)$ is undefined for initial contexts.

Example 3. Fig. 3 illustrates parts of the symbolic execution of the ATM example TIOSTS given in Fig. 1.

So far, we have defined the symbolic execution of a TIOSTS without any adjustments related to our testing concerns from TIOSTS. In the following, we will complete the symbolic execution with quiescent configurations, i.e., identifying situations where the system can remain silent. The system is usually expected to react by sending messages when it receives a message from its environment. However, sometimes, it cannot emit an output from any given state [4,15,18,25,28]. In such a case, the inability of the system to react becomes a piece of information. To make this fact clear we enrich symbolic execution by adding a special output action $\delta!$ to denote the absence of output in those specific deadlock situations.

Definition 3 (Quiescence enrichment). *The quiescence enrichment* $SE(\mathbb{G})^\delta$ *of* $SE(\mathbb{G}) = (ec_0, \mathbb{EC})$ *is* $(ec_0, \mathbb{EC}^\delta)$ *where* $\mathbb{EC}^\delta$ *is the set* $\mathbb{EC}$ *enriched by new execution contexts* ec^δ. *For each context* $ec = (q, \pi, \lambda, ev, pec)$ *in* $\mathbb{EC}$, *a new context* $ec^\delta = (q, \pi \wedge \pi^\delta, \lambda, (f^{dur}(ec), \delta!), ec)$ *is considered where*[1]:

$$\pi^\delta = \bigwedge_{\substack{pec(ec')=ec \\ chan(ec') \in C^{out}}} \left(\forall f^{dur}(ec). \forall f^{out}_{chan(ec')}(ec). \neg \pi(ec') \right)$$

Let us emphasize that π^δ is satisfiable only for contexts ec where there is no choice of values for the variables for triggering from ec a transition carrying an emission. The context could not be considered quiescent if such a choice were possible, i.e. if there would exist an output transition towards an execution context ec', for which there is an instantiation of variables $f^{dur}(ec)$ and $f^{out}_{chan(ec')}(ec)$ making the condition $\pi(ec')$ true.

Example 4. We discuss some examples from Fig. 3. The execution context ec_0 does not have successors with outputs (π^δ is *True*) which denotes that the ATM is awaiting withdrawal requests or non-involved debit authorizations. This quiescent situation is captured by adding the context $ec_0^\delta = (q_0, True, \lambda(ec_0), (\mathbf{z_0}, \delta!), ec_0)$. The execution context ec_1 has three successors with outputs, ec_2, ec_9 and ec_{10}. Then, π^δ is:

$$\bigwedge_{j \in \{2,9,10\}} \forall f^{dur}(ec_1). \forall f^{out}_{chan(ec_j)}(ec_1). \neg \pi(ec_j)$$

which is not satisfiable. Thus, there is no need to add a quiescent transition from ec_1. The same applies to ec_2 and ec_3.

A symbolic path of $SE(\mathbb{G})^\delta = (ec_0, \mathbb{EC}^\delta)$ is a sequence $p = ec_0.ec_1 \ldots ec_n$ where ec_0 is the initial context, for $i \in [1, n]$, $ec_i \in \mathbb{EC}^\delta$, and $pec(ec_i) = ec_{i-1}$. $Paths(SE(\mathbb{G})^\delta)$ denotes the set of all such paths. We will use the notation $tgt(p)$

---

[1] with the convention that $\bigwedge$ quantified over empty conditions is the formula *True*.

to refer to ec_n, the last context of p. We define the set of traces of a symbolic path p in $Paths(SE(\mathbb{G})^\delta)$ by:

$$Traces(p) = \bigcup_{\nu \in M^F} \{\ \nu(p) \mid M \models_\nu \exists F^{ini}.\pi(tgt(p))\ \}$$

where ν applies to a path p of the form $p'.ec$ as $\nu(p) = \nu(p').\nu(ev(ec))$ with the convention $\nu(ev(ec_0)) = \epsilon$ and $\nu(ev(ec)) = (\nu(z), c?\nu(x))$ (resp. $(\nu(z), c!\nu(y))$ or $(\nu(z), \delta!)$) if $ev(ec)$ is of the form $(z, c?x)$ (resp. $(z, c!y)$ or $(z, \delta!)$).

By solving the path condition of a given path, we can evaluate all symbolic events occurring in the path and extract the corresponding trace. The set of traces of $\mathbb{G}$ is defined as :

$$Traces(\mathbb{G}) = \bigcup_{p \in Paths(SE(\mathbb{G})^\delta)} Traces(p)$$

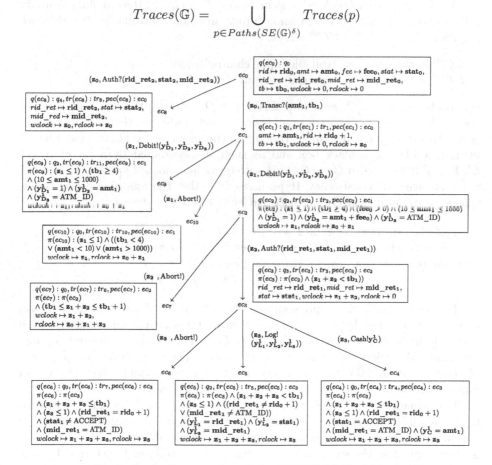

Fig. 3. Symbolic execution of the ATM TIOSTS of Fig. 1

4   Conformance Testing

Conformance testing aims to check that a system under test behaves correctly w.r.t a reference model, a TIOSTS $\mathbb{G}$ in our case. The test case stimulates the system with inputs and observes the system's outputs, their temporalities, and the quiescent situations to compare them to those specified by $\mathbb{G}$. For generality, we propose test cases that control some inputs of the system under test while leaving other inputs driven by systems in its environment. We assume that the test case selects inputs and observes outputs on some channels while it can only observe inputs and outputs on other channels of the system under test. Illustrating with the ATM, a test case provides the ATM with withdrawal requests, and observes withdrawal authorizations received from the bank.

We characterize a *Localized System Under Test (LUT)* (terminology from [6, 14]) tested in a context where some inputs are not controllable. For this, we partition $C^{in} = CC^{in} \amalg UC^{in}$ where:

- CC^{in} is the set of controllable input channels and
- UC^{in} is the set of uncontrollable input channels.

For a set of channels C, we denote $Evt(C) = D \times Act(C)$ the set of all concrete events that can occur in a trace: an event (d, act) indicates that the occurrence of action act happens d units of time after the previous event. In model-based testing, a LUT is a black box and as such, is abstracted by a set of traces $LUT \subseteq Evt(C_\delta)^*$ with $C_\delta = C \cup \{\delta\}$ satisfying additional hypotheses, denoted as $\mathcal{H}$, ensuring its consistency. Hypotheses $\mathcal{H}$ gather the following 3 properties, where for σ_1, σ_2 in $Evt(C)^*$, d in D and $ev \in Evt(C)$ we have:

- *stable by prefix:* $\sigma_1.\sigma_2 \in LUT \Rightarrow \sigma_1 \in LUT$
- *quiescence:* for any $d < delay(ev)$, $\sigma_1.ev \in LUT \Rightarrow \sigma_1.(d, \delta!) \in LUT$
- *input complete:* for any $d < delay(ev)$, $c \in CC^{in}$, $v \in M$,
$$\sigma_1.ev \in LUT \Rightarrow \sigma_1.(d, c?v) \in LUT$$

The first hypothesis simply states that every prefix of a system trace is also a system trace. The hypothesis on quiescence states that if an event ev whose action is in $act(C)$ occurs in LUT, then LUT can remain quiescent for any duration strictly less than the delay of the event. The hypothesis on input completeness enables LUT to receive any input on a controllable channel, i.e., an input received from the test case, during the delay of any ev in the LUT.

The semantics of a TIOSTS $\mathbb{G}$, denoted by $Sem(\mathbb{G})$, will include traces with the admissible temporary observation of quiescence: if an event $ev = (d, act)$ is specified in $\mathbb{G}$ then quiescence can be observed for any duration $d' < d$. $Sem(\mathbb{G})$ is then defined as the smallest set containing $Traces(\mathbb{G})$ and such that for any $\sigma \in Evt(C)^*$, $ev \in Evt(C)$, for any $d < delay(ev)$:
$$\sigma.ev \in Traces(\mathbb{G}) \Rightarrow \sigma.(d, \delta!) \in Sem(\mathbb{G})$$

As other previous works [23, 29] have already done to suit their needs, we are now slightly adapting the conformance relation of [21]:

Definition 4 (tioco). *Let C be a set of channels. Let $\mathbb{G}$ and LUT be resp. a TIOSTS defined on C and a subset of $Evt(C_\delta)^*$ satisfying $\mathcal{H}$.*
LUT tioco $\mathbb{G}$ iff for all $\sigma \in Sem(\mathbb{G})$, for any $ev \in Evt(C^{out} \cup \{\delta\})$, we have:

$$\sigma.ev \in LUT \Rightarrow \sigma.ev \in Sem(\mathbb{G})$$

The relation *tioco* states that LUT is in conformance with $\mathbb{G}$, if and only if after a specified sequence σ observed on LUT, any event produced by LUT as an output or a delay of quiescence, leads to a sequence $\sigma.ev$ of $sem(\mathbb{G})$.

Test case generation is often based on the selection of a test purpose which permits to choose a particular behavior in $\mathbb{G}$ to be tested [4,8,15,17]. As symbolic execution plays a key role both for the semantics of TIOSTS and for testing issues in general, whether it is for the test case generation or the verdict computation, our test purposes will be paths $tp \in Paths(SE(\mathbb{G})_\delta)$ with satisfiable path conditions, i.e. verifying $Sat(tgt(tp))$. As outputs are involved in the *tioco* relation, we require tp to end with an output event, i.e., $chan(tgt(tp)) \in C^{out}$.

Contrary to [15], our test purposes are simple paths and not (finite) subtrees, simplifying the construction of test cases. We will not need to consider the case where the observed behavior on LUT corresponds to several symbolic paths simultaneously. By taking it a step further, to avoid such tricky situations completely, we restrict ourselves to symbolic paths that do not induce nondeterminism. In line with [3,19], we forbid that there are two outgoing transitions of an execution context concerning the same channel which can be covered by the same trace. Unlike [3,19] which impose determinism conditions at the state level, we deal with uninitialized variables at the path level:

Definition 5 (Test purpose). *Let $tp \in Paths(SE(\mathbb{G})^\delta)$ be a symbolic path verifying $Sat(tgt(tp))$ and $chan(tgt(tp)) \in C^{out}$. Let $\mathbb{EC}(tp)$ be its set of execution contexts.*
tp is a test purpose *for $\mathbb{G}$ if tp satisfies the so-called* trace-determinism property*: for ec in $\mathbb{EC}(tp)$ and ec' in $\mathbb{EC}(\mathbb{G})$ s.t. $Sat(ec')$, $pec(ec) = pec(ec')$, $tr(ec) \neq tr(ec')$, and $chan(ec) = chan(ec')$, the following formula is unsatisfiable:*

$$\left(\exists F^{ini}.\pi(ec)\right) \bigwedge \left(\exists F^{ini}.\pi(ec')\right)$$

The trace-determinism property simply expresses that from any intermediate execution context of tp, it is impossible to deviate in $\mathbb{G}$ with a common trace, independently of the initial conditions.

Example 5. The test purpose $tp = ec_0.ec_1 \ldots ec_4$ given in Fig. 3 satisfies tracedeterminism. To support our comments, let us consider a simpler TIOSTS with three transitions (tr_1, tr_2 and tr_3), with two of them, tr_2 and tr_3, creating a non-deterministic situation:
$tr_1 : q_0 \xrightarrow{\text{Transc?}amt} q_1, tr_2 : q_1 \xrightarrow{\text{Debit!}amt} q_2$ and $tr_3 : q_1 \xrightarrow{[\,fee>0\,],\,\text{Debit!}amt+fee} q_3$
The TIOSTS symbolic execution for some initial execution context ec_0 ($F^{ini} = \{amt_0, fee_0\}$) can reach execution contexts ec_1 ($tr(ec_1) = tr_1$, $pec(ec_1) = ec_0$), ec_2 ($tr(ec_2) = tr_2$, $pec(ec_2) = ec_1$), and ec_3 ($tr(ec_3) = tr_3$, $pec(ec_3) = ec_1$),

building 2 symbolic paths $ec_0.ec_1.ec_2$ and $ec_0.ec_1.ec_3$. Respective path conditions are $\pi(ec_2) = (\mathbf{y_D^1} = \mathbf{amt_1})$ and $\pi(ec_3) = (\mathbf{fee_0} > 0) \wedge (\mathbf{y_D^1} = \mathbf{amt_1} + \mathbf{fee_0})$ where $\mathbf{amt_1}$ binds the value received on the channel Transc $(f_{\text{Transc}}^{in}(ec_0) = \mathbf{amt_1})$ and $\mathbf{y_D^1}$ binds the value emitted on the channel Debit $(f_{\text{Debit}}^{out}(ec_1) = \mathbf{y_D^1})$.

Given $tp = ec_0.ec_1.ec_2$, execution contexts ec_2 and ec_3 share the same output channel Debit and the same predecessor context ec_1. tp satisfies the trace-determinism property. Indeed, the formula:
$$\left(\exists\{\mathbf{fee_0}, \mathbf{amt_0}\}.(\mathbf{y_D^1} = \mathbf{amt_1})\right) \bigwedge \left(\exists\{\mathbf{fee_0}, \mathbf{amt_0}\}.(\mathbf{fee_0} > 0) \wedge (\mathbf{y_D^1} = \mathbf{amt_1} + \mathbf{fee_0})\right)$$
is not satisfiable because $\mathbf{amt_1} < \mathbf{amt_1} + \mathbf{fee_0}$ holds as we have $\mathbf{fee_0} > 0$. A trace cannot belong to distinct paths: if the debit value is the same as what is requested for withdrawal then the trace covers ec_2, else ec_3 is covered.

5 Path-Guided Test Cases

Roughly speaking, a test case $\mathbb{TC}$ will be a mirror TIOSTS of a TIOSTS $\mathbb{G}$, restricted by tp, a test purpose of $\mathbb{G}$, intended to interact with a *LUT* that we wish to check its conformance to $\mathbb{G}$ up to tp. $\mathbb{TC}$ will be a tree-like TIOSTS with tp of $\mathbb{G}$ as a backbone, incorporating the following specific characteristics:

– execution contexts in $\mathbb{EC}(tp)$ constitute the main branch,
– sink states or leaves are assimilated with a test verdict. Notably, the last execution context $tgt(tp)$ of tp will be assimilated with the PASS verdict,
– from each ec in $\mathbb{EC}(tp)$ other than $tgt(tp)$, the outgoing arcs outside tp decline all ways to deviate from tp and directly lead to a verdict state, either an inconclusive verdict or a failure verdict. In Definition 6, we will specify the different ways of constructing these arcs from tp states, leading to a verdict.

In Definition 6, we define $\mathbb{TC}$ by enumerating the different cases of transitions to be built according to channel type and tp membership. We now give a few indications for enhancing the readability.

– Channel roles are reversed: channels in CC^{in} as well as channel δ (resp. $C^{out} \cup UC^{in}$) become output (resp. input) channels;
– Variables of $\mathbb{TC}$ will be symbolic variables involved in tp and will be used to store successive concrete events observed on *LUT*;
– Any observation on *LUT* will be encoded as an input transition in $\mathbb{TC}$, whether it is an emission from a channel of C^{out}, a reception on a channel of UC^{in} or a time-out observation.

On the latter, it is conventional to consider that a system that does not react before a certain delay, chosen to be long enough, is in a state of quiescence. The only notable exception is when input transitions of tp give rise to output transitions for $\mathbb{TC}$, modeling a situation in which $\mathbb{TC}$ stimulates *LUT* by sending it data. The choice of the data to send is conditioned by two constraints: (i) taking into account the information collected so far and stored in the variables of $\mathbb{TC}$ in the first steps, and (ii) the guarantee of being able to reach the last

EC of tp, i.e. the verdict PASS. This will be done by ensuring the satisfiability of the path condition of tp, leaving aside the variables already binded.

We will refer to variables of a symbolic path as follows: for[2] $\alpha \in \{in, out\}$, $f^\alpha(p) = \bigcup_{i \in [1,n)} f^\alpha(ec_i)$ will denote all introduced fresh variables in F^α used to compute a symbolic path $p = ec_0.ec_1 \ldots ec_n$. Similarly, $f^{dur}(p) = \{f^{dur}(ec_i)|i \in [1, n)\}$. Moreover, for a target execution context $ec_n = tgt(p)$, we denote $\overline{f}^\alpha(ec_n)$ and $\overline{f}(ec_n)$ resp. for $f^\alpha(p)$ and $f(p)$.

Definition 6 (Path-guided test case). *Let tp be a test purpose for a TIOSTS $\mathbb{G}$. Let us consider the signature $\widehat{\Sigma} = (\widehat{A}, \widehat{K}, \widehat{C})$ where:*

- *$\widehat{A} = f^{in}(tp) \cup f^{out}(tp)$,*
- *$\widehat{K} = f^{dur}(tp)$,*
- *$\widehat{C} = C_\delta$ such that $\widehat{C}^{in} = C^{out} \cup UC^{in}$ and $\widehat{C}^{out} = CC^{in} \cup \{\delta\}$*

A test case guided by tp is a TIOSTS $\mathbb{TC} = (\widehat{Q}, \widehat{q_0}, \widehat{Tr})$ over $\widehat{\Sigma}$ where:

- *$\widehat{Q} = (\mathbb{EC}(tp) \setminus \{tgt(tp)\}) \cup \mathbb{V}$ where:*
  $\mathbb{V} = \{$ PASS, $FAIL^{out}$, $FAIL^{dur}$, INC^{out}, INC^{dur}, INC^{ucIn}_{spec}, INC^{ucIn}_{uspec} $\}$,
- *$\widehat{q_0} = ec_0$,*

- *$\widehat{Tr}$ is defined by a set $\mathcal{R}$ of 10 rules of the form $\dfrac{H \quad (Ri)}{tr \in \widehat{Tr} \quad \text{LABEL}}$ for $i \in [1, 10]$.*
  Such a rule reads as follows: the transition tr is added due to rule Ri to $\widehat{Tr}$ provided that hypothesis H holds and $Sat(\phi(tr))$.

In writing the rules of $\mathcal{R}$, we will use the following formulas

ϕ_{stim} : $\exists F^{ini} \cup \overline{f}(tgt(tp)) \setminus \overline{f}(ec').\pi(tp)$

ϕ^{obs}_{spec} : $f^{dur}(ec) < \text{TM} \wedge (\exists F^{ini}.\pi(ec'))$
ϕ^{obs}_{uspec} : $f^{dur}(ec) < \text{TM} \wedge \bigwedge_{\substack{pec(ec')=ec \\ chan(ec')=c}} (\forall F^{ini}. \neg \pi(ec'))$

ϕ^{δ}_{spec} : $f^{dur}(ec) \geq \text{TM} \wedge \bigvee_{\substack{pec(ec')=ec \\ chan(ec') \in C^{out} \cup UC^{in} \cup \{\delta\}}} (\exists F^{ini}.\pi(ec'))$
ϕ^{δ}_{uspec} : $f^{dur}(ec) \geq \text{TM} \wedge \bigwedge_{\substack{pec(ec')=ec \\ chan(ec') \in C^{out} \cup UC^{in} \cup \{\delta\}}} (\forall F^{ini}. \neg \pi(ec'))$

where the constant TM (Time-out) sets the maximum waiting-time for observing outputs or uncontrollable inputs.

$$\dfrac{ec \xrightarrow{(z,c?x)} ec' \in \mathbb{EC}(tp) \quad c \in CC^{in}}{(ec, c!x, \phi_{stim}, \{f^{dur}(ec)\}, id_{\widehat{A}}, ec') \in \widehat{Tr}} \text{ SKIP} \quad (R1)$$

$$\dfrac{\begin{array}{cc} ec \xrightarrow{(z,c!y)} ec' \in \mathbb{EC}(tp) \\ ec' \neq tgt(tp) \quad c \in C^{out} \end{array}}{(ec, c?y, \phi^{obs}_{spec}, \{f^{dur}(ec)\}, id_{\widehat{A}}, ec') \in \widehat{Tr}} \text{ SKIP} \quad (R2) \qquad \dfrac{\begin{array}{cc} ec \xrightarrow{(z,c!y)} ec' \in \mathbb{EC}(tp) \\ ec' = tgt(tp) \quad c \in C^{out} \end{array}}{(ec, c?y, \phi^{obs}_{spec}, \emptyset, id_{\widehat{A}}, \text{PASS}) \in \widehat{Tr}} \text{ PASS} \quad (R3)$$

---

[2] For i and j in $\mathbb{N}$ verifying $i < j$, $[i, j)$ contains the integers from i to $j-1$ included.

$$\frac{ec \xrightarrow{(z,c!y)} ec' \quad ec \in \mathbb{EC}(tp) \setminus \{tgt(tp)\}}{ec' \notin \mathbb{EC}(tp) \quad c \in C^{out}} \quad (R4)}{(ec, c?y, \phi_{spec}^{obs}, \emptyset, id_{\widehat{A}}, \text{INC}^{out}) \in \widehat{Tr}} \text{INC}^{out} \qquad \frac{ec \in \mathbb{EC}(tp) \quad c \in C^{out}}{(ec, c?f_c^{out}(ec), \phi_{uspec}^{obs}, \emptyset, id_{\widehat{A}}, \text{FAIL}^{out}) \in \widehat{Tr}} \quad (R5)}{\text{FAIL}^{out}}$$

$$\frac{ec \xrightarrow{(z,c?x)} ec' \in \mathbb{EC}(tp)}{c \in UC^{in}} \quad (R6)}{(ec, c?x, \phi_{spec}^{obs}, \{f^{dur}(ec')\}, id_{\widehat{A}}, ec') \in \widehat{Tr}} \text{SKIP} \qquad \frac{ec \xrightarrow{(z,c?x)} ec' \quad ec \in \mathbb{EC}(tp) \setminus \{tgt(tp)\}}{ec' \notin \mathbb{EC}(tp) \quad c \in UC^{in}} \quad (R7)}{(ec, c?x, \phi_{spec}^{obs}, \emptyset, id_{\widehat{A}}, \text{INC}_{spec}^{ucIn}) \in \widehat{Tr}} \text{INC}_{spec}^{ucIn}$$

$$\frac{ec \in \mathbb{EC}(tp) \quad c \in UC^{in}}{(ec, c?f_c^{in}(ec), \phi_{uspec}^{obs}, \emptyset, id_{\widehat{A}}, \text{INC}_{uspec}^{ucIn}) \in \widehat{Tr}} \quad (R8)}{\text{INC}_{uspec}^{ucIn}}$$

$$\frac{ec \in \mathbb{EC}(tp)}{(ec, \delta!, \phi_{spec}^{\delta}, \emptyset, id_{\widehat{A}}, \text{INC}^{dur}) \in \widehat{Tr}} \quad (R9)}{\text{INC}^{dur}} \qquad \frac{ec \in \mathbb{EC}(tp)}{(ec, \delta!, \phi_{uspec}^{\delta}, \emptyset, id_{\widehat{A}}, \text{FAIL}^{dur}) \in \widehat{Tr}} \quad (R10)}{\text{FAIL}^{dur}}$$

A verdict PASS is reached when tp is covered, verdicts INC_n^m are reached when traces deviate from tp while remaining in $\mathbb{G}$, and verdicts FAIL^m denote traces outside $\mathbb{G}$. The annotations n and m provide additional information on the cause of the verdict. Rules $R1$, $R2$ and $R6$, grouped together under the label SKIP, allow advancing along tp, resp. by stimulating LUT with the sending of data, observing an emission on C^{out} and observing a reception on UC^{in}. Rule $R3$ indicates that the last EC of tp, thus the PASS verdict, has been reached. Rules $R4$, $R7$, $R8$ and $R9$, each with a label INC_n^m indicate that the observed event causes LUT to leave tp, without leaving $\mathbb{G}$, resp. by observing an output, an input specified in $\mathbb{G}$, an input not specified in $\mathbb{G}$ and a time-out observation. Lastly, rules $R5$ and $R10$, resp. labeled by FAIL^{out} and FAIL^{dur}, raise a FAIL verdict, for resp. an unauthorized output and an exceeded time-out.

Example 6. In Fig. 4, the test case for $tp = ec_0.ec_1 \ldots ec_4$ (see Example 5) is depicted. Certain verdict states are repeated for readability, and defining rules annotate the transitions. The test case utilizes the Transc channel as a controllable input channel for stimulation while observing all other channels. For space considerations, we comment only on some rules.

Rule $R1$ defines a stimulation action $\text{Transc}!(\mathbf{amt_1}, \mathbf{tb_1})$ (transition from ec_0 to ec_1) constrained by $\pi(tp)$ to select an appropriate value for $\mathbf{amt_1}$ and $\mathbf{tb_1}$ together with a time of stimulation $\mathbf{z_0}$ that allows to follow the test purpose. Within $\pi(tp)$, $\mathbf{amt_1}$ is limited to some range ($10 \leq \mathbf{amt_1} \leq 1000$), $\mathbf{tb_1}$ is greater than or equal 4, while $\mathbf{z_0}$ is unconstrained (a duration measured with clock $\mathbf{z_0}$). Non-revealed variables, i.e., other than $\mathbf{z_0}$, $\mathbf{amt_1}$ and $\mathbf{tb_1}$ appearing in $\pi(tp)$ are bound by the existential quantifier due to their unknown values at this execution point. The clock $\mathbf{z_1}$ is reset to enable reasoning on subsequent actions' duration (measured on $\mathbf{z_1}$). Rule $R2$ defines an observation action $\text{Debit}?(\mathbf{y_{D_1}^1}, \mathbf{y_{D_2}^1}, \mathbf{y_{D_3}^1})$ constrained by $(\mathbf{z_1} < \text{TM}) \wedge \exists \mathbf{fee_0}.\pi(ec_2)$ (transition from ec_1 to ec_2) while rule $R5$ defines the same observation constrained by $(\mathbf{z_1} < \text{TM}) \wedge \forall \mathbf{fee_0}.\neg \pi(ec_2) \wedge$

$\forall\{\mathbf{fee_0}\}.\neg\pi(ec_9)$ (transition from ec_1 to FAIL^{out}). Both situations are possible: Trace $(0,\mathrm{Transc!}(50,4)).(0,\mathrm{Debit?}(1,51,\mathrm{ATM\_ID}))$ reaches ec_2 whereas FAIL_{out} is reached by traces $(0,\mathrm{Transc!}(50,4)).(0,\mathrm{Debit?}(1,0,\mathrm{ATM\_ID}))$ and $(0,\mathrm{Transc!}(50,4)).(2,\mathrm{Debit?}(1,51,\mathrm{ATM\_ID}))$ due resp. to data and time non-compliance. Rule $R6$ defines an unspecified quiescence $\delta!$ constrained by $(\mathbf{z_1}\geq \mathrm{TM})\wedge\forall\mathbf{fee_0}.\neg\pi(ec_2)\wedge\forall\mathbf{fee_0}.\neg\pi(ec_9)\wedge\forall\mathbf{fee_0}.\neg\pi(ec_{10})$ (transition from ec_1 to FAIL^{dur}). The trace $(0,\mathrm{Transc!}(50,4)).(5,\delta!)$ reaches FAIL^{dur} (time-out TM is set to 5): the time-out is exceeded without the mandatory output on the channel Debit being observed. Rule $R9$ defines a specified quiescence $\delta!$ (transition from ec_1 to INC^{dur} not drawn for space). This case arises when there is still sufficient time to reach ec_2, ec_9 or ec_{10} (no quiescence applies from ec_1, see Example 4).

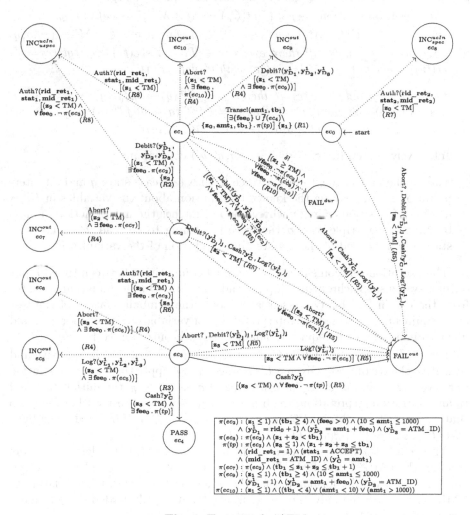

Fig. 4. Test case for ATM

A test case $\mathbb{TC}$ interacts with a *LUT*, designed to comply with a TIOSTS $\mathbb{G}$, to issue a verdict about a test purpose *tp*. $\mathbb{TC}$ is therefore defined as a mirror image of $\mathbb{G}$: emissions (receptions) in $\mathbb{TC}$ correspond to receptions (emissions) of $\mathbb{G}$ (cf. rules $R1$ and $R2$), except uncontrollable channels whose actions are not reversed (cf rule $R6$). Given a concrete action *act* (with v for value received or sent), we denote $\overline{act}$ its mirror action, defined as follows: $\overline{c!v} = c?v$ for $c \in C^{out}$, $\overline{c?v} = c!v$ for $c \in CC^{in}$ and $\overline{c?v} = c?v$ for $c \in UC^{in}$.

We introduce an execution relation that abstracts a synchronized execution of a trace with $\mathbb{TC}$:

Definition 7 (Relation execution $\leadsto_{\mathbb{TC}}$). *Let $\mathbb{G}$ be a TIOSTS and tp a test purpose for $\mathbb{G}$ with $\mathbb{TC} = (\widehat{Q}, \widehat{q_0}, \widehat{Tr})$ the test case guided by tp.*

The execution relation *$\leadsto_{\mathbb{TC}} \subseteq \left(Evt(C_\delta)^* \times \widehat{Q} \times M^F \right)^2$ is defined by:*

for $ev.\sigma \in Evt(C_\delta)^*$, $q, q' \in \widehat{Q}$ and for $\nu, \nu' \in M^F$, $(ev.\sigma, q, \nu) \leadsto_{\mathbb{TC}} (\sigma, q', \nu')$ holds iff there exists $tr \in \widehat{Tr}$ s.t. $src(tr) = q$, $tgt(tr) = q'$, $\nu'(act(tr)) = act(ev)$ and $M \models_{\nu'} \phi(tr)$ with ν' defined as:

- *if $ev = (d, c?v)$ and $chan(tr) = c$ then $\nu' = \nu[f^{dur}(q) \mapsto d][f_c^{in}(q) \mapsto v]$;*
- *if $ev = (d, c!v)$ and $chan(tr) = c$ then $\nu' = \nu[f^{dur}(q) \mapsto d][f_c^{out}(q) \mapsto v]$;*
- *else, i.e., $ev = (d, \delta!)$ and $chan(tr) = \delta$, $\nu' = \nu[f^{dur}(q) \mapsto d]$.*

Intuitively, a step $(ev.\sigma, q, \nu) \leadsto_{\mathbb{TC}} (\sigma, q', \nu')$ consists in:

- reading the first element ev of a trace from a test case state q and an interpretation ν synthesizing the known information about the variables in F;
- finding a transition tr in $\widehat{Tr}$ whose action matches the mirror action of ev;
- building a new triple with σ the trace remaining to be read, q' a successor state of q in $\widehat{Q}$, and ν' the updated interpretation of the variables F.

The execution relation simulates a parallel composition between timed input output systems, synchronizing inputs and outputs. Our formulation deviates from the one in [22] for two essential reasons: the symbolic nature of the test case requires the intermediate interpretations of variables to be memorized, and uncontrollable channels require to adapt the synchronization [11].

Let *LUT* be a subset of $Evt(C_\delta)^*$ satisfying $\mathcal{H}$ and $\overset{*}{\leadsto}_{\mathbb{TC}}$ be the reflexive and transitive closure of $\leadsto_{\mathbb{TC}}$. Given a *LUT* trace σ_0, we apply the execution relation iteratively from an initial triplet consisting of σ_0 a trace, $\widehat{q_0}$ the initial state and ν_0 an arbitrary interpretation, to obtain the corresponding test verdict for *tp*, so that the verdict set obtained from the execution of $\mathbb{TC}$ on *LUT* is defined by: $vdt(LUT, \mathbb{TC}) = \{V \mid \exists \sigma_0 \in LUT, \nu_0 \in M^F, (\sigma_0, \widehat{q_0}, \nu_0) \overset{*}{\leadsto}_{\mathbb{TC}} (\sigma, V, \nu)\}$.

Theorem 1 states the soundness of the test case execution for detecting errors through the $FAIL^{out}$ and $FAIL^{dur}$ verdicts. A proof can be found in [5]. Comparable results can be formulated for the other verdicts. Still, those relating to FAIL verdicts are the only ones to guarantee that any discarded system under test does not satisfy the tioco conformance relation.

Theorem 1. *Let C be a set of channels. Let $\mathbb{G}$ and LUT be resp. a TIOSTS defined on C and a subset of $Evt(C_\delta)^*$ satisfying $\mathcal{H}$.*
If LUT tioco $\mathbb{G}$ then for any test purpose tp for $\mathbb{G}$ with $\mathbb{TC}$ as test case guided by tp, we have $\mathrm{FAIL}^{out} \notin vdt(LUT, \mathbb{TC})$ and $\mathrm{FAIL}^{dur} \notin vdt(LUT, \mathbb{TC})$.

The test case generation is implemented as a module in the Diversity symbolic execution platform [9]. Resulting test cases are expressed in Diversity's entry language, allowing their exploration through symbolic execution with the SMT-solver Z3 [10]. For easier execution, we export the test cases from Diversity in JSON format, with transition guards expressed in the SMT-LIB input format for SMT-solvers. Our experiments involved applying this test case generation to the ATM example on an Intel Core i7 processor. Varying the size of the test purposes up to 100 transitions, we observed successful trace-determinism verification for all test purposes. We noted a noticeable increase in generation duration as the test purpose size grew while still remaining feasible. Generating the TIOSTS test case (513 transitions) for the test purpose of 100 transitions took more than 40s, in contrast to only 500 ms for the test purpose of size 4 in Example 6 (31 transitions) resp. 8s for the test purpose of size 50 (138 transitions).

6   Related Work

Existing works for (t)ioco conformance test cases from symbolic models employs two main generation methods: online and offline. Online generation [8,12,15] involves dynamically generating test cases while exploring the model during execution on the system under test. In contrast, offline generation [3,4,16] focuses on deriving test cases from the model before executing them on the system under test. Some works [8,12,15] propose online test case generation using symbolic execution. Yet, these works did not consider time constraints, and in particular, work [8] did not consider quiescence. In [8], a test purpose is a finite path, while in [15], it is a finite symbolic subtree. Both works require maintaining a set of reached symbolic states during test case execution to avoid inconsistent verdicts in case of non-determinism, at the expense of computational resources for tracking the symbolic states and solving their path conditions. Work [4] proposes offline test case generation using a path to compute a timed stimulation sequence for the system under test. The recorded timed output sequence is then analyzed for conformance. This approach lacks control over the value and timing of the next stimulation relative to the observed system behavior, potentially resulting in greater deviations from the test purpose. In [16], objective-centered testers for timed automata are built using game theory. Works [3,19] propose offline test case generation as tree-like symbolic transition systems and thus restricted to determinism as we do. The test case generation in [19] relies on abstract interpretation to reinforce test case guards on data to keep chances of staying in the test purpose and does not consider time. In [3], symbolic execution techniques are used for data handling, while zone-based abstraction techniques are employed for time. This separation results in less expressive and flexible models, as it cannot extend to incorporate data parameters in the time constraints.

7 Conclusion

This paper presents an offline approach to conformance test case generation from models of timed symbolic transition systems using symbolic execution to handle data and time. Our test purpose is a symbolic path in the model that fulfills a determinism condition to enable the generation of sound tree-like test cases. By distinguishing between controllable inputs (from the test case) and uncontrollable inputs (from other systems), our approach enhances the usability of test cases when the system interacts with other systems (remote in general). This allows our test cases to be used in more liberal configurations, typically those that may appear for distributed systems. It's worth noting that our test cases include configurations where the resolution time exceeds the stimulation time or overlaps the arrival of an observation. These points will be the subject of future work. Similarly, the experiments described in this paper concern the computation of test cases derived from models that have not yet been used to test systems. This is another avenue for future work.

References

1. Aichernig, B.K., Tappler, M.: Symbolic input-output conformance checking for model-based mutation testing. In: USE@FM 2015, Elsevier (2015). https://doi.org/10.1016/j.entcs.2016.01.002
2. Alur, R., Dill, D.L.: A theory of timed automata. Theor. Comput. Sci. (1994). https://doi.org/10.1016/0304-3975(94)90010-8
3. Andrade, W.L., Machado, P.D.L., Jéron, T., Marchand, H.: Abstracting time and data for conformance testing of real-time systems. In: ICST Workshops (2011). https://doi.org/10.1109/ICSTW.2011.82
4. Bannour, B., Escobedo, J.P., Gaston, C., Le Gall, P.: Off-line test case generation for timed symbolic model-based conformance testing. In: ICTSS (2012). https://doi.org/10.1007/978-3-642-34691-0_10
5. Bannour, B., Lapitre, A., Le Gall, P., Nguyen, T.: Symbolic path-guided test cases for models with data and time, version of this paper extended with appendix (2023). https://doi.org/10.48550/arXiv.2309.06840
6. Benharrat, N., Gaston, C., Hierons, R.M., Lapitre, A., Le Gall, P.: Constraint-based oracles for timed distributed systems. In: Yevtushenko, N., Cavalli, A.R., Yenigün, H. (eds.) ICTSS 2017. LNCS, vol. 10533, pp. 276–292. Springer, Cham (2017). https://doi.org/10.1007/978-3-319-67549-7_17
7. de Boer, F.S., Bonsangue, M.: On the nature of symbolic execution. In: ter Beek, M.H., McIver, A., Oliveira, J.N. (eds.) FM 2019. LNCS, vol. 11800, pp. 64–80. Springer, Cham (2019). https://doi.org/10.1007/978-3-030-30942-8_6
8. van den Bos, P., Tretmans, J.: Coverage-based testing with symbolic transition systems. In: Beyer, D., Keller, C. (eds.) TAP 2019. LNCS, vol. 11823, pp. 64–82. Springer, Cham (2019). https://doi.org/10.1007/978-3-030-31157-5_5
9. CEA: diversity, eclipse formal modeling project. https://projects.eclipse.org/proposals/eclipse-formal-modeling-project (2023)
10. de Moura, L., Bjørner, N.: Z3: an efficient SMT solver. In: Ramakrishnan, C.R., Rehof, J. (eds.) TACAS 2008. LNCS, vol. 4963, pp. 337–340. Springer, Heidelberg (2008). https://doi.org/10.1007/978-3-540-78800-3_24

11. Escobedo, J.P., Gaston, C., Gall, P.L., Cavalli, A.R.: Testing web service orchestrators in context: a symbolic approach. In: SEFM, pp. 257–267. IEEE Computer Society (2010)
12. Frantzen, L., Tretmans, J., Willemse, T.A.C.: Test generation based on symbolic specifications. In: FATES (2004). https://doi.org/10.1007/978-3-540-31848-4_1
13. Frantzen, L., Tretmans, J., Willemse, T.A.C.: A symbolic framework for model-based testing. In: Havelund, K., Núñez, M., Roşu, G., Wolff, B. (eds.) FATES/RV -2006. LNCS, vol. 4262, pp. 40–54. Springer, Heidelberg (2006). https://doi.org/10.1007/11940197_3
14. Gaston, C., Hierons, R.M., Le Gall, P.: An implementation relation and test framework for timed distributed systems. In: ICTSS (2013). https://doi.org/10.1007/978-3-642-41707-8_6
15. Gaston, C., Le Gall, P., Rapin, N., Touil, A.: Symbolic execution techniques for test purpose definition. In: TestCom (2006). https://doi.org/10.1007/11754008_1
16. Henry, L., Jéron, T., Markey, N.: Control strategies for off-line testing of timed systems. Formal Methods Syst. Des. (2022). https://doi.org/10.1007/s10703-022-00403-w
17. Hessel, A., Larsen, K.G., Mikucionis, M., Nielsen, B., Pettersson, P., Skou, A.: Testing real-time systems using UPPAAL. In: FORTEST (2008). https://doi.org/10.1007/978-3-540-78917-8_3
18. Janssen, R., Tretmans, J.: Matching implementations to specifications: the corner cases of ioco. In: Hung, C., Papadopoulos, G.A. (eds.) SAC. ACM (2019). https://doi.org/10.1145/3297280.3297496
19. Jéron, T.: Symbolic model-based test selection. In: Machado, P.D.L. (ed.) SBMF. Elsevier (2008). https://doi.org/10.1016/j.entcs.2009.05.051
20. King, J.C.: Symbolic execution and program testing. Commun. ACM **19**(7), 385–394 (1976)
21. Krichen, M., Tripakis, S.: Black-box conformance testing for real-time systems. In: Graf, S., Mounier, L. (eds.) SPIN 2004. LNCS, vol. 2989, pp. 109–126. Springer, Heidelberg (2004). https://doi.org/10.1007/978-3-540-24732-6_8
22. Krichen, M., Tripakis, S.: Interesting properties of the real-time conformance relation tioco. In: Barkaoui, K., Cavalcanti, A., Cerone, A. (eds.) ICTAC 2006. LNCS, vol. 4281, pp. 317–331. Springer, Heidelberg (2006). https://doi.org/10.1007/11921240_22
23. Luthmann, L., Göttmann, H., Lochau, M.: Compositional liveness-preserving conformance testing of timed I/O automata. In: Arbab, F., Jongmans, S.-S. (eds.) FACS 2019. LNCS, vol. 12018, pp. 147–169. Springer, Cham (2020). https://doi.org/10.1007/978-3-030-40914-2_8
24. Marsso, L., Mateescu, R., Serwe, W.: TESTOR: a modular tool for on-the-fly conformance test case generation. In: Beyer, D., Huisman, M. (eds.) TACAS 2018. LNCS, vol. 10806, pp. 211–228. Springer, Cham (2018). https://doi.org/10.1007/978-3-319-89963-3_13
25. Rusu, V., Marchand, H., Jéron, T.: Automatic verification and conformance testing for validating safety properties of reactive systems. In: Fitzgerald, J., Hayes, I.J., Tarlecki, A. (eds.) FM 2005. LNCS, vol. 3582, pp. 189–204. Springer, Heidelberg (2005). https://doi.org/10.1007/11526841_14
26. Standard, E.: Methods for testing and specification (MTS); the testing and test control notation version 3; Part 1: TTCN-3 core language (2005)

27. von Styp, S., Bohnenkamp, H., Schmaltz, J.: A conformance testing relation for symbolic timed automata. In: Chatterjee, K., Henzinger, T.A. (eds.) FORMATS 2010. LNCS, vol. 6246, pp. 243–255. Springer, Heidelberg (2010). https://doi.org/10.1007/978-3-642-15297-9_19
28. Tretmans, J.: Test generation with inputs, outputs, and quiescence. In: Margaria, T., Steffen, B. (eds.) TACAS 1996. LNCS, vol. 1055, pp. 127–146. Springer, Heidelberg (1996). https://doi.org/10.1007/3-540-61042-1_42
29. Tretmans, J., Janssen, R.: Goodbye ioco. In: a journey from process algebra via timed automata to model learning. Springer, Cham (2022). https://doi.org/10.1007/978-3-031-15629-8_26

Model-Based Testing of Asynchronously Communicating Distributed Controllers

Bence Graics(✉), Milán Mondok, Vince Molnár, and István Majzik

Department of Measurement and Information Systems, Budapest University of Technology and Economics, Műegyetem rkp. 3., 1111 Budapest, Hungary
{graics,molnarv,majzik}@mit.bme.hu

Abstract. Programmable controllers are gaining prevalence even in distributed safety-critical infrastructures, e.g., in the railway and aerospace industries. Such systems are generally integrated using multiple loosely-coupled reactive components and must satisfy various critical requirements. Thus, their systematic testing is an essential task, which can be encumbered by their distributed characteristics. This paper presents a model-based integration test generation approach leveraging hidden formal methods based on the collaborating statechart models of the components. Statecharts can be integrated using various composition modes (e.g., synchronous and asynchronous) and then transformed (via a symbolic transition systems formalism – XSTS) into the input formalisms of model checker back-ends, namely UPPAAL, Theta and Spin in an automated way. The model checkers are utilized for test generation based on multiple coverage criteria. The approach is implemented in our open source Gamma Statechart Composition Framework and evaluated on industrial-scale distributed controller subsystems from the railway industry.

Keywords: Model-based integration testing · Collaborating statecharts · Asynchronous communication · Hidden formal methods · Tool suite

1 Introduction

Software-intensive programmable controllers are becoming increasingly widespread in safety-critical infrastructure, including railway interlocking systems (RIS) [25,32] and onboard computers of satellites [1,29,41]. Such systems are commonly integrated using multiple components that may communicate in different ways, e.g., synchronously or asynchronously. This way, these components form a loosely coupled distributed architecture. In general, these systems are embedded into their environments and system components must cooperate to conduct complex tasks in response to external commands and environmental or internal events (e.g., changes in the context or component failures) to reach their objectives. Thus, they are often referred to as *reactive systems*.

© The Author(s), under exclusive license to Springer Nature Switzerland AG 2024
J. Cámara and S.-S. Jongmans (Eds.): FACS 2023, LNCS 14485, pp. 23–44, 2024.
https://doi.org/10.1007/978-3-031-52183-6_2

The distributed architecture of these systems may complicate their development, necessitating precise means to describe the functional behavior of the components, as well as their integration, including their execution (e.g., sequential, concurrent or parallel) and communication modes (e.g., signal- or message-based). In addition, as these systems carry out safety-critical tasks, testing their implementation in a systematic way is vital. Nonetheless, the verification of component interactions is often encumbered by the imprecise, informal definition and interpretation of their execution and communication modes.

To aid the development of complex distributed systems, model-based and component-based systems engineering (MBSE and CBSE) [3,21,27,39] approaches advocate the application of reusable models and components based on high-level modeling languages, e.g., SysML and UML. These modeling languages allow for the platform-independent description of functional behavior in terms of components, e.g., statecharts [19] for reactive behavior, as well as structure, e.g., block diagrams to define the (hierarchical) integration of system components.

In turn, most MBSE and CBSE approaches do not provide refined and extensible tool-centric means for the automated and systematic testing of the system implementation [30]. These insufficiencies may stem from informal model semantics or the lack of sound and efficient verification methods, e.g., the lack of integration between the high-level models and verification back-ends [9,31].

Consequently, verification-oriented MBSE and CBSE approaches for developing distributed control systems should support – in an integrated tool suite – i) high-level modeling languages with *precise semantics* to describe the behavior of standalone components, and *component integration* (composition) based on different modes of execution and communication, ii) automated *model transformations* with *traceability* that map these models into different *model checkers* to support the *exhaustive verification* of functional behavior in a flexible and reusable way, and iii) *test generation* algorithms suitable for large-scale systems based on various *coverage criteria* to test the system implementation efficiently.

To address these challenges, we propose a fully automated model-based testing (MBT) approach in our open source Gamma Statechart Composition Framework[1] [34]. At its core, the approach builds on a high-level *statechart language* (GSL) [12] and a *composition language* (GCL) [17] with precise semantics to describe the functional behavior of standalone components, as well as their integration using various execution and communication modes (jointly, referred to as *composition* modes) to support the modeling of synchronous and asynchronous systems. The emergent models are automatically mapped into a low-level formalism, called *EXtended Symbolic Transition Systems* (XSTS), serving as a *common formal representation* to capture reactive behavior. *Formal verification* based on temporal properties is supported by mapping XSTS models and properties into the input formalisms of different model checker back-ends; so far, UPPAAL [4] and Theta [40] have been supported. The mappings feature *model reduction*

---

[1] More information about the framework (e.g., preprints) and the source code can be found at http://gamma.inf.mit.bme.hu/ and https://github.com/ftsrg/gamma/.

and *model slicing* algorithms to allow for the verification of large-scale systems. Back-annotation facilities automatically map the verification results into a high-level *trace language* (GTL) [17]. Building on the formal verification and back-annotation functionalities [10], the approach generates *integration tests* based on customizable structural (model element-based), dataflow-based and behavior-based (interactional) coverage criteria. Test cases are *optimized* and concretized to different platforms, e.g., C or Java, to detect faults in the implementation of components (e.g., missing implementation of states or transitions), incorrect variable definitions and uses or component interactions.

Our previous works [16, 17, 34] focused on the design and verification of *timed* systems (with UPPAAL) and reactive systems featuring *signal-* and *shared variable* based communication (with Theta). The already presented version of the framework featured GSL and GCL, the model transformations into the UPPAAL and Theta model checker back-ends along with the model reduction and slicing algorithms, and automated test generation based on the coverage criteria. This work extends the framework to support *asynchronous* systems with *message-based* communication and integrates the open source Spin [23, 24] model checker, which is tailored to verifying models with such characteristics. The integration necessitates i) a mapping between the framework's behavioral representation (XSTS) and Promela (the input language of Spin) and ii) the validation of our approach, including the evaluation of its performance. The results show that Spin provides additional versatility and thus, is a valuable addition to the framework.

The novel contributions of the paper are as follows:

1. the *EXtended Symbolic Transition Systems* (XSTS) formalism (see Sect. 4) as a *general representation* for reactive components and their integration that supports high-level control structures (see their application in Sect. 5) to facilitate efficient verification;
2. a nontrivial *mapping* between the XSTS and Promela languages (see Sect. 6) integrating the open source Spin model checker into Gamma, including two message queue representation modes for asynchronous communication using i) *arrays* and ii) native *asynchronous channels* in Promela; and
3. the *evaluation* of our extended approach on real-life distributed controller subsystems from the railway industry, comparing the UPPAAL, Theta and Spin integrated model checkers and their underlying algorithms (see Sect. 7).

2   Related Work

The main features of our approach revolve around i) a low-level analysis language (XSTS) to describe reactive behavior and allow for formal verification using model checkers and ii) an end-to-end integrated MBT approach for composite high-level engineering models using hidden formal methods (i.e., users do not have to familiarize themselves with the underlying analysis models and algorithms). Thus, we present related work according to these aspects.

Analysis Languages. BoogiePL [6], as the input formalism of the Boogie model checker, is a generic intermediate language for verifying object-oriented programs. The language is coarsely typed and offers constructs such as procedures and arrays. The main distinguishing feature of BoogiePL compared to XSTS is that it was specifically tailored to fit *computer programs*, not reactive behavior, and offers constructs for this specific domain (like procedures).

PVS [36] is a specification and verification language based on classical, typed higher-order logic and can be used for *theorem proving*, a lower-level approach to construct and maintain large formalizations and logical proofs.

The Symbolic Analysis Laboratory (SAL) [38] language is an intermediate language tailored to *concurrent transition systems*. It is intended to be the target for translators that extract the behavior descriptions from other languages, and a common source for different analysis tools. It is supported by the SAL toolset that includes symbolic (BDD-based) and bounded (SAT/SMT-based) model checkers. In turn, SAL provides limited support for traceability and back-annotation with respect to the integration of high-level modeling languages.

In comparison, we offer XSTS, an extension of *symbolic transition systems* close to the input formalisms of model checkers, which also includes *control structures* to retain several characteristics of the engineering models that can be utilized in mappings for model checking (e.g., parallel execution), and also supports easy traceability and back-annotation regarding the verification results.

Integrated Test Generation Approaches. The idea of using *hidden formal methods* to *generate tests* based on *integrated models* has been applied in several tool-based approaches. The CompleteTest tool [8] analyzes software written in the Function Block Diagram (FBD) language. The approach generates tests based on logical coverage criteria (e.g., MC/DC) by mapping FBD into UPPAAL.

Literature [33] introduces the AutoMOTGen toolset that allows for the mapping of Simulink/Stateflow models into the SAL framework and the model checking based test generation based on different logical coverage criteria.

AGEDIS [20] is an MBT toolset for component-based distributed systems. It integrates model and test suite editors, test simulation and debugging tools, test coverage and defect analysis tools, and report generators. Test models can be defined using UML class, state machine and object diagrams. Test generation uses the so-called TGV engine [26] and supports state and transition coverage.

Smartesting CertifyIt [28] is also an MBT toolset, which integrates editors for defining requirements and traceability, test adapters and test models. Test models are composed of UML class diagrams (for data description), state machines and object diagrams (initial states of executions), as well as Business Process Model and Notations (BPMN). Tests are generated with the CertifyIt Model Checker, which supports state, transition and transition-pair coverage criteria.

Our Gamma framework is unique from the aspect that it supports various composition modes [17] (including support for asynchronous communication), formally defined but configurable test coverage criteria with model reduction and slicing algorithms [16], and model checkers integrated via the XSTS formalism.

3   Extended Model-Based Testing Approach

This section first overviews the underlying *modeling languages* that support our extended model-based testing (MBT) approach for component-based reactive systems. Next, the section details the *steps* (user activities and automated internal model transformations) that constitute the i) model design ii) formal verification and iii) test generation phases of our approach in Gamma.

- The **Gamma Statechart Language (GSL)** is a UML/SysML-inspired *configurable* formal statechart [19] language supporting different semantic variants of statecharts, e.g., different kinds of priorities for transitions [12].
- The **Gamma Composition Language (GCL)** is a composition language for the formal hierarchical composition of state-based (GSL) components according to multiple *execution* and *communication* (composition) modes [17]. It supports synchronous systems where components communicate with *sampled signals* and are executed *concurrently* (*synchronous-reactive*) or *sequentially* (*cascade*), as well as asynchronous systems where components communicate with *queued messages* and are executed *sequentially* (*scheduled asynchronous-reactive* [14]) or *in parallel* (*asynchronous-reactive*).
- The **Gamma Genmodel Language (GGL)** is a configuration language for configuring model transformations, e.g., to select the model checker for verification or the coverage criteria for test generation.
- The **Gamma Property Language (GPL)** is a property language supporting the definition of CTL* [7] properties and thus, the formal specification of requirements regarding (composite) component behavior.
- The **EXtended Symbolic Transition Systems (XSTS)** (Sect. 4) is a low-level formalism tailored to formal verification that supports the description of reactive behavior based on a set of *variables*, the values of which represent system states, and *transitions* that describe the possible state changes.
- The **Gamma Trace Language (GTL)** is a high-level trace language for reactive systems, supporting the description of execution traces, i.e., reached state configurations, variable values and output events in response to input events, time lapse and scheduling from the environment [17]. Such execution traces are also interpretable as *abstract tests* as the language also supports the specification of general assertions targeting the values of variables.

Figure 1 depicts the modeling languages, modeling artifacts and the model transformations of our MBT approach in Gamma. In the following, we overview the steps of the workflow, i.e., *model design, formal verification* and *test generation*; for more details regarding the formalization of the *test coverage criteria* and *model reduction* and *slicing* algorithms, we direct the reader to [16].

Model Design. The model design phase consists of three steps. As an optional step, *external component models*, i.e., statecharts created (Step 0) in integrated modeling tools (front-ends), are *imported* to Gamma by model transformations (Step 1) that map these models into GSL statecharts. Currently, the import of Yakindu [15], MagicDraw, SCXML [37], and XSML [14] models are supported.

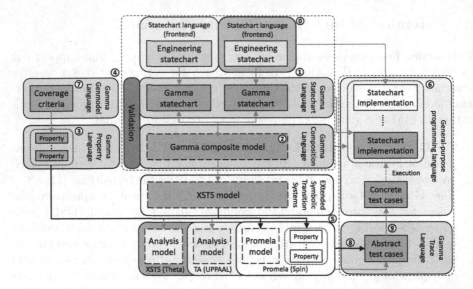

Fig. 1. Modeling languages and transformation chains of our approach in Gamma.

Next, the GSL models can be *hierarchically integrated* in GCL (Step 2) according to different (and potentially mixed) execution and communication modes [17].

Formal Verification. The integrated GCL model can be *formally verified* using a sequence of *automated model transformations* that map the model and verifiable properties (specified in GPL – Step 3) into inputs of *model checker* back-ends via XSTS, i.e., UPPAAL, Theta or the newly integrated Spin. The mappings are configured using GGL (Step 4) to select back-ends, add optional constraints (e.g., scheduling) and set reduction and slicing algorithms [16]. The selected model checker *exhaustively explores* the model's state space with respect to the given property (Step 5) and potentially returns a diagnostic trace that is *back-annotated* to the GCL model, creating a representation in GTL.

Test Generation. If the verification results are satisfactory, *implementation* from the models can be derived manually or automatically using the code generators of Gamma or the integrated modeling front-ends (Step 6). *Integration tests* for the implementation can be *generated* based on the GCL model using the aforementioned formal verification facilities [10]. Test generation is driven by customizable dataflow-based, structural (model element-based) and behavior-based (interactional) *coverage criteria* (Step 7), which are *formalized* as reachability properties in GPL representing *test targets* (see [16] for their formalization), and control the model checkers to compute *test target (criterion) covering paths* during model traversal; model checking time can be limited by *timeouts* to discard uncoverable test targets. These paths, i.e., returned witnesses for satisfying the reachability properties, are represented as GTL execution traces (Step 8) and, in a testing context, are regarded as *abstract test cases* for the property based on which they are generated. *Optimization algorithms* are used to i) prevent starting model checking runs for already covered test targets and ii) remove unnecessary tests that do not contribute to the coverage of the specified criteria [16]. The abstract tests are concretized (Step 9) to execution environments (e.g., JUnit) by generating concrete calls to provide test inputs, time delay and schedule system execution, and then retrieve and evaluate outputs to check the *conformance* of the system model and implementation for these particular traces.

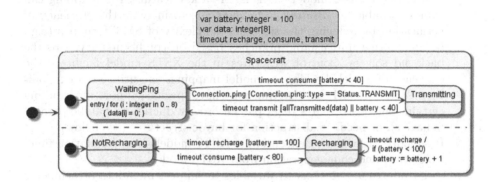

Fig. 2. Excerpt of the *spacecraft* component model.

For example, if we aim to cover with tests each *state node* in a GCL model, we can use the *state-coverage* keyword (with potential *include* and *exclude* constraints regarding the targeted state nodes) in a GGL *analysis model transformation* task besides selecting the model checker back-end for test generation. As a result, GPL reachability properties are generated for every targeted state node (in the form of E F state $regionName.stateNodeName$) besides the analysis model, which are, one by one, translated into the input property formalism of the selected model checker and checked on the generated analysis model. In case the targeted state node is reachable, the model checker returns a *diagnostic trace* consisting of a series of steps (alternating sequence of states and input events) leading to its coverage. This diagnostic trace is automatically back-annotated to GTL, creating an *abstract test case* that serves as a basis for test concretization.

Note that the architecture of the framework supports the *parallel running* of multiple model checkers (model checker portfolio) for test generation, this way, tackling the problem of finding the most suitable back-end for a given model.

4   EXtended Symbolic Transition Systems

The *symbolic transition systems* (STS) formalism [22] is a commonly used low-level representation, e.g., for hardware model checking. STS models consist of two SMT [2] formulas that describe the *set* of *initial states* and the *transition relation*. Our novel *EXtended Symbolic Transition Systems* (XSTS) formalism introduced here is built on top of STS, and serves as a formal representation for the behavior of component-based reactive systems.

The key contributions of our XSTS formalism compared to STS are as follows:

– XSTS introduces an *imperative layer* on top of SMT formulas, i.e., a set of *control structures* with operational semantics, having two advantages:
  1. It allows for the direct mapping of XSTS models into STS models, and thus, into SMT formulas (see the Appendix for details) for their verification by SMT-based model checkers. As a key feature, the mapping can exploit variable *substitution techniques* [35] to increase the efficiency of verification by reducing the size and complexity of SMT formulas (e.g., by eliminating unreachable branches based on conditions) given to the back-end solvers. Control structures in the XSTS model facilitate the efficiency of model checking, as model mappings can utilize the capabilities of target model checkers that may have specific support for handling certain control structures (e.g., partial order reduction to handle parallel behavior), or peculiar abstraction for havoc structures (see below).
  2. It supports the concrete execution of XSTS models (i.e., simulation functionalities) and code generation from XSTS models.
– XSTS introduces *annotations* to variables to support metadata attachments (e.g., to identify control variables). Annotations support *traceability* between source and target models (e.g., GCL and derived XSTS models) that can be leveraged during verification (e.g., special handling of the control variables).
– XSTS introduces *clock* variables to support time-dependent behavior.
– XSTS *partitions* the transition relation of STS to allow for distinguishing between the behavior of the environment and that of the modeled system. This feature can be leveraged during verification (back-annotation of diagnostic traces to higher-level models) and also during test generation by separating the steps of the environment and that of the modeled system in a generated execution trace, which is important when mapping to concrete test calls.

In the following, we introduce the elements of the XSTS language based on an example *spacecraft* statechart model (see Fig. 2) originally defined in [16] that transmits *data* to a *ground station* while managing its *battery*. Note that in general, the XSTS language has all the features necessary to capture in a semantic-preserving way the behavior of integrated GSL and GCL models.

Type Declarations. An XSTS model (see the left snippet of Fig. 3) begins with *custom type declarations* (**type** keyword) that contain *literals*, similarly to enum types in programming languages.

Variable Declarations. Type declarations are followed by global *variable declarations* (**var** keyword) with *integer*, *boolean*, *clock*, and the previously discussed *custom* types. The language also supports *array* types, which are mathematical SMT arrays, similar to the *map* data structure of programming languages (see Line 12). Variable declarations can optionally contain an *initial value* (Line 8).

Variables annotated with the **ctrl** keyword (Lines 6–7) are *control variables*, indicating that these variables contain *control* information, which can be exploited during verification (e.g., in Theta when using different abstraction algorithms).

Transitions. Model behavior is defined by three *transitions*, which are *atomic*, i.e., they are either *executed* in their *entirety* or *not at all*. The system's internal behavior is described by the **trans** transition (Lines 14–44), while the behavior of the system's environment is described by the **env** transition (Lines 48–58). The **init** transition (Lines 45–47) initializes the system. Regarding their execution order, the **init** transition is executed first, after which the **env** and **trans** transitions alternate. In our example, the **init** transition sets the statechart's initial state, while the **env** transition places a random message (more precisely, its identifier) in its queue. The **trans** transition pops a message from the queue, which is processed in both regions (see Sect. 6 for the details of queue handling).

Basic Statements. The detailed behavior of transitions is captured via *statements*. *Assign* statements (see Line 19) assign a value of its domain to a single variable. *Assume* statements (Line 34) act as guards; they can be executed only if their condition holds. *Havoc* statements assign a nondeterministically selected value of its domain to a variable (Line 55). *Local variable declarations* can be used to create *transient* variables that are only accessible in the scope they were created in and are not part of the system's state vector (Line 54).

Composite Statements. Composite statements contain other statements (operands), and can be used to describe complex control structures. *Sequences* are lists of statements that are executed sequentially; each statement operates on the result of the previous statement. *Choice* statements (see Line 33) model non-deterministic choices between multiple statements; only one branch is selected for execution, which cannot contain failing assumptions, i.e., if every branch contains failing assumptions, then the choice statement also fails. *Parallel* statements support the parallel execution of the operands (Lines 24–44). *If-else* statements are deterministic choices based on a condition (see Lines 25–31) with an optional *else* branch. XSTS also supports deterministic *for* loops over ranges (Line 30).

5   Transforming Asynchronous GCL Models into XSTS

The XSTS-Promela mapping (to be presented in Sect. 6) exploits the traits of the GCL-XSTS transformation (see the transformations related to Steps 4 and 5 in Sect. 3). Thus, we first overview the relevant parts of this automated transformation before moving onto the XSTS-Promela mapping in the next section.

Synchronous statechart components (defined in GSL) are the basic building blocks of GCL models (see Step 2 in Sect. 3). Asynchronous components are created from simple or composite synchronous ones (i.e., synchronous-reactive or cascade composite components) by wrapping them using a so-called *asynchronous adapter* [17]: an adapter maps *signals* used in the synchronous domain

```
1   type Status : { TRANSMIT, ... }
2   type Communication : { _Inact_,
        WaitingPing, Transmitting }
3   type Battery : { _Inact_, ... }
4   var ping : boolean = false
5   var ping_types : Status = TRANSMIT
6   ctrl var com : Communication = _Inact_
7   ctrl var bat : Battery = _Inact_
8   var batteryLevel : integer = 0
9   var recharge, transmit, consume : clock
10  // Message queue related variables
11  @QueueSize var len : integer
12  @Queue var msgQueue : [integer] -> integer
13  @Queue var argQueue : [integer] -> Status
14  trans {
15    if (len > 0) {
16      local var id: integer := msgQueue[0];
17      msgQueue[0] := 0; // Message pop
18      if (id == 1) {
19        ping := true;
20        ping_types := argQueue[0];
21        argQueue[0] := 0; // Arg. pop
22      } ... // Potentially other msgIds
23      len := len - 1;
24      par {
25        if (com == WaitingPing && ping &&
            ping_types == TRANSMIT) {
26          com := Transmitting;
27        } else if (... // Second transition
28        else if (com == Transmitting && ... <=
            transmit) {
29          com := WaitingPing;
30          for i from 0 to 7 do { data[i] := 0; }
31        }
32      } and {
33        choice {
34          assume (bat == NotRecharging && ... <=
              consume && batteryLevel < 80);
35          bat := Recharging;
36          recharge := 0;
37        } or {... // Second transition
38        } or {
39          assume (bat == Recharging && ... <=
              recharge && batteryLevel == 100);
40          bat := NotRecharging;
41        } or {
42          assume !(...); // Else branch
43      } } ... // Input event clearing
44    } }
45  init { ... // Variable initializations
46    com := WaitingPing; // Initial sates
47    bat := NotRecharging; ... } // Entries
48  env {
49    if (len <= 0) {
50      local var msgId : integer;
51      havoc msgId;
52      if (0 < msgId && msgId <= 1) {
53        msgQueue[len] := msgId;
54        local var argument : Status;
55        havoc argument;
56        argQueue[len] := argument;
57        len := len + 1; }
58  } }
```

```
1   mtype:Status = { Status_TRANSMIT, ... }
2   mtype:Communication = { C_Inact_,
        WaitingPing, Transmitting }
3   ... // Proc-sync related variables
4   chan chan_parallel_0 = [0] of { bit };
5   chan chan_parallel_1 = [0] of { bit };
6   // Metadata variables
7   byte flag = 0; bit isStable = 0; //
8   proctype Parallel_0() {
9     if
10    :: (com == WaitingPing && ping &&
          ping_types == Status_TRANSMIT) -> com
          = Transmitting;
11    :: else -> if
12    ... // Second transition
13      :: else -> if
14        :: (com == Transmitting && ... <=
            transmit) -> com = WaitingPing;
15          for (i : 0 .. 7) { data[i] = 0; }
16        :: else
17        fi;
18      fi;
19    fi;
20    chan_parallel_0 ! 1; // Finished execution
21  }
22  proctype Parallel_1() {
23    if
24    :: if :: (bat == NotRecharging &&
          batteryLevel < 80); fi;
25      bat = Recharging;
26      recharge = 0;
27    :: if :: ... // Second transition
28    :: if :: (bat == Recharging && ... <=
          recharge && batteryLevel == 100); fi;
29      bat = NotRecharging;
30    :: else;
31    fi;
32    chan_parallel_1 ! 1; // Finished execution
33  }
34  proctype EnvTrans() { (flag > 0);
35    ENV: isStable = 1; // Verification may end
36    atomic {
37      isStable = 0;
38      if // Mapping a havoc statement
39      :: msgId = 0;
40      :: msgId = 1;
41      fi;
42      if :: (0 < msgId && msgId <= 1) ->
43      ... // Mapping-specific array handling
44      fi;
45    }
46    TRANS: atomic { flag = 2; // TRANS is next
47      run Parallel_0(); run Parallel_1();
48      bit msg_parallel_0 = 0; // Process sync
49      chan_parallel_0 ? msg_parallel_0;
50      chan_parallel_1 ? msg_parallel_0;
51      ... // Input event clearing
52      flag = 1; }; // ENV is next
53    goto ENV;
54  }
55  init { ... // Variable initializations
56    { run EnvTrans(); flag = 1; } // Run proc.
57  }
```

Fig. 3. XSTS and Promela representations of the *spacecraft* component of Fig. 2.

into *messages* and related *message queues* in the asynchronous domain and vice versa, i.e., the two representations of *events* in the two domains.

The transformation uses the same steps for every asynchronous adapter (detailed in [13]). Here, we overview the relevant parts of this procedure related to asynchronous communication based on the standalone *spacecraft* component and its (automatically generated) XSTS representation described in Fig. 3.

As for GSL statecharts, the XSTS elements corresponding to the *transitions* are created (Lines 24–43), e.g., *parallel* statements corresponding to orthogonal regions, containing *if-else* structures and *choice* statements corresponding to the transitions of state nodes (considering hierarchy) in the particular regions. Note that *literals* of a *custom type declaration* stored in a *control variable* correspond to the state nodes of a region.

As for composite (GCL) synchronous components, their execution and interactions are handled by (hierarchically) composing the created XSTS statements corresponding to the contained components, and defining shared variable based communication (corresponding to signal transmission) using *assign* statements.

```
1   int queue[Capac]; // Queue array
2   int len = 0; // Size variable
3   // Nonempty
4   len > 0
5   // Append
6   queue[len] = value; len = len + 1;
7   // Peek
8   int peekVariable = queue[0];
9   ... // Referencing peekVariable
10  peekVariable = 0; // Resetting "local" var
11  // Pop
12  queue[0] = queue[1]; queue[1] = queue[2];..
13  queue[len - 1] = 0;
14  len = len - 1;
```

```
1   chan queue = [Capac] of { int }; // Queue
2   // Nonempty
3   len(queue) > 0
4   // Append
5   queue ! value;
6   // Peek
7   int peekVariable;
8   queue ? <peekVariable>;
9   ... // Referencing peekVariable
10  peekVariable = 0; // Resetting "local" var
11  // Pop
12  int popVariable;
13  queue ? popVariable;
14  popVariable = 0; // Resetting "local" var
```

Fig. 4. XSTS-Promela mapping's two supported message queue representation modes.

For the asynchronous adapters defined in GCL, the related XSTS constructs "wrap" the statements representing synchronous behavior by introducing and handling additional variables related to the representation of message queues (see Lines 15–23). For a single message queue in an asynchronous adapter, the transformation introduces one or more annotated *array* variables (denoting that the variable represents a queue) depending on whether the queue stores only non-parameterized messages (*msgQueue* array, Line 12) or also message parameter values, i.e., payload (one or more *argQueue* arrays for every *parameter type* to enable storing each parameter value of each message; Line 13). In addition, an integer variable with an annotation is introduced (*len*, see Line 11) storing the number of messages present in the queue (the annotation denotes that the variable stores the size of a queue). In general, every message type stored in the queue is assigned an integer identifier that is appended to the *msgQueue* array, modeling the append of the message instance to the corresponding message queue (see Line 53). In case the message is parameterized, the parameter value is appended to the corresponding *argQueue* array (see Line 56).

Asynchronous message handling is defined using the *nonempty*, *peek* and *pop* message queue operations (the Promela representation of these operations is summarized in the left snippet of Fig. 4). An *if* statement is created that checks whether the *msgQueue* variable corresponding to the message queue is *nonempty* (*len* > 0 condition, see Line 15), and if so, the stored message identifier is retrieved (*peek* and *pop*, see Lines 16–17) and based on it, the corresponding input event (signal) variable is set to true (Line 19). Potentially, the parameter values are also loaded from the *argQueues* to the corresponding input parameter variables (*peek* and *pop*, see Lines 20–21). Finally, the constructs representing the synchronous behavior are wrapped into the created *if* statement.

Regarding the verification of the emergent XSTS models, the different model checkers (i.e., corresponding mappings) integrated into Gamma support slightly different subsets of the XSTS language, as well as different property languages, i.e., supported subsets of the GPL language (see Table 1): UPPAAL fully supports the verification of time-dependent behavior with a restricted CTL property language, whereas Spin fully supports the verification of parallel behavior (see *parallel* statements in Sect. 6) and LTL [7]; in turn, Theta has experimental support for these features and supports only reachability properties. Nonetheless, every model checker supports the XSTS constructs used for asynchronous communication as presented in this section, as well as the specification of reachability properties to capture test targets in our MBT approach.

Table 1. Overview of features of the mappings and underlying model checkers supported by our MBT approach in Gamma. ✓ = full support; ✗ = experimental

Model checker (back-end)	Model representation	Parallel behavior	Timed behavior	Asynchronous communication	Property language
Theta	XSTS	✗	✗	✓	Reachability
UPPAAL	Timed automata		✓	✓	Restricted CTL
Spin	Promela process models	✓		✓	LTL

6   Mapping XSTS Models into Promela

The integration of Spin into the Gamma framework relies on a semantic-preserving mapping between XSTS and Promela. The mapping distinguishes XSTS elements (see Sect. 4) that i) *do have* or ii) *do not* have a *direct semantic equivalent* in Promela, as well as iii) constructs related to *asynchronous communication*. The only XSTS element whose semantic-preserving mapping is not supported is *clock variable*, as Promela does not support time-dependent behavior: for such variables, the mapping introduces integer variables in the Promela model and "discretizes" component execution according to user-defined constraints.

In the case of i), the mapping is simple and creates a single Promela element with the same semantics, as presented in Table 2. Note that several constructs are optimized, e.g., the mapping resets local variable declarations at the end of their declaring scope and uses *d_step* for the sequences of assignments. Contrarily, in the case of ii) and iii), multiple Promela elements are created to preserve the semantics of the original XSTS elements and asynchronous communication.

In the following, we consider XSTS constructs related to ii) and iii). First, we present the mapping of XSTS *transitions*, *havoc* and *parallel* statements based on the *spacecraft* model represented in Fig. 3. Then, we describe asynchronous communication supported by two message queue representation modes.

Table 2. XSTS elements and their *direct semantic equivalent* element(s) in Promela.

XSTS	Promela
boolean, integer, custom, array type	**bit, int, mtype, array** type
(*local*) **variable declaration**	(*local*) **variable declaration** (+ *resets*)
assume statement	**boolean** expression (in **if** construct)
assignment statement	**assignment** statement
sequence (of statements)	**sequence** (of statements) (+ *d_step*)
choice statement	*nondeterministic multiary* **if** construct
if-else statement	*binary* **if** construct with **else**
loop statement	**for** deterministic iteration statement

Mapping Elements Without a Direct Semantic Equivalent. The *init* transition is mapped into Promela's *init* process (see Lines 55–57 in the Promela code), which, after executing the mapped initialization statements, runs the *EnvTrans* process (Line 34). The *EnvTrans* process comprises (in a *cycle*) the statements of the *env* and *trans* transitions; the corresponding statements are wrapped in *atomic* blocks and labeled *ENV* (Line 35) and *TRANS* (Line 46). The states of execution are indicated by the *isStable* and *flag* (meta)variables (Line 7): the former is true iff *trans* has finished, but the execution of *env* has not started yet (needed to specify valid *end states* in property specifications – Line 35), while the latter encodes which transition is under execution: 0 – *init*, 1 – *env* and 2 – *trans* (needed for back-annotation – Lines 46 and 52).

Havoc statements describe nondeterministic assignments from the targeted variable's domain and thus, they are mapped into nondeterministic multiary *if* selection constructs whose *options* describe the possible values based on the variable's type (see Line 38). For *boolean* variables, *true* and *false* (0 and 1) values, for *custom* types, the declared *literals* are included. *Integer* variables are handled only if their domain is *restricted*, e.g., in the case of stored message identifiers; otherwise this construct is not supported in the Promela mapping.

Parallel statements (i.e., their operands; see Lines 8 and 22) are mapped into Promela processes and synchronization constructs. First, the *local variables* (if any) referenced from the operands are identified, which are mapped into *parameter declarations* in the corresponding processes (*assignments* to local variables are not supported). Next, *synchronization channels* are created (Lines 4–5) to allow for synchronization between the caller and the parallelly running processes: the caller waits for the started processes to finish execution (Lines 47–50).

Mapping Elements of Asynchronous Communication. As for asynchronous communication, the XSTS-Promela mapping builds on the traits of the GCL-XSTS transformation (see Sect. 5), in particular, the annotated array variables (*msgQueue* and *argQueue*) and *len* variables, jointly representing the message queues of asynchronous adapters. As a configuration option, the mapping features two (fixed capacity) message queue representation modes based on i) *arrays* and ii) native *asynchronous channels* in Promela. Note that the former can be considered as a straightforward mapping (resembling the one to UPPAAL), whereas the latter serves as an experiment in the context of Promela, potentially being more efficient (see Sect. 7). Figure 4 summarizes how the *nonempty*, *append*, *peek*, *pop* and *size* queue handling operations are represented in the two message queue representation modes. Note that Fig. 3 uses the array-based mode, but the other version could be easily reproduced based on Fig. 4.

In the former case, the mapping is straightforward: the XSTS array variables and *len* variables are mapped into Promela array variables and integer variables. The statements that refer to these variables and correspond to message queue operations are also mapped according to the rules presented in Table 2.

The latter mode uses the *asynchronous channel* construct of Promela that natively supports the *size* queue operation using the built-in *len* function, disregarding the XSTS integer variable *len*. To support the *nonempty*, *append*, *peek* and *pop* message queue operations, the mapping traverses the XSTS model and identifies the corresponding statements (i.e., the constructs of the left snippet of Fig. 4) based on variable annotations and pattern matching, which are mapped to native message queue related constructs in Promela (right snippet of Fig. 4).

7   Practical Evaluation

This section evaluates our extended MBT approach, focusing on the features of the model mappings via the XSTS formalism and their performance in the context of the integrated model checker back-ends, Theta, UPPAAL and Spin.

We conduct this evaluation in alignment with the needs of our industrial partner that develops railway control systems in the context of customized MBSE and CBSE approaches. During development, our partner must conform to safety standards (cf. EN 50128 [5]), which require (among others) integration test generation based on these models to check the system implementation. Thus, our partner has interest in conducting this task in an automated and efficient way.

In regard to test generation, we already showed the feasibility of our approach for synchronous models in [16] using the UPPAAL and Theta model checkers. Nevertheless, our partner uses different composition semantics at different hierarchy levels of a system model (considering different deployment modes), and is interested in generating tests for components with asynchronous communication, too. Accordingly, our partner needs information regarding the characteristics of the supported model checkers and message queue representations.

Thus, we formulated the following research questions (RQ) for our evaluation: how efficient are the integrated model checkers of Gamma for *test generation*, in terms of *generation time* and *generated test set size*, on

RQ-1. *synchronous* models (*shared variable* based communication), and

RQ-2. *asynchronous* models (*message queue* based communication) using a i) native *channel* based (relevant only in the case of Spin), and ii) *array*-based mapping of GCL/XSTS message queues?

Models. We evaluated the RQ on two system models received from our industrial partner, namely *railway signaller subsystem* (RSS) and *railway interlocking system* (RIS), each corresponding to the characteristics of a specific RQ.[2]

RSS comprises the model of a subsystem used in railway traffic control systems [11]. It builds on statechart components (two *antivalence checkers* connected to a *signaller*) communicating in a synchronous (signal-based) way with many interaction points (dispatch/reception of signals in different states). Thus, it is a relevant model for generating tests that check interactions. RSS consists of 26 state nodes, 97 transitions and 5 variables.

The RIS model [11] defines an industrial communication protocol used in RIS, and comprises three components defined in the proprietary XSML language (integrated to Gamma in [14]), namely *control center*, *dispatcher* and *object handler*. The components are executed sequentially, and communicate with their messages stored in local message queues; the *control center* and *object handler* can communicate only via the *dispatcher*. Thus, testing interactions based on messages besides covering state nodes and transitions is relevant in this model's context. The model contains 38 state nodes, 118 transitions and 23 variables.

Measurement Settings. We used different *composition modes* in the different models to capture the components' expected execution and communication modes. For the RSS model, we used the *cascade* and *synchronous-reactive* composition modes [17]. For RIS, we used the *scheduled asynchronous-reactive* mode [14] and the capacity of 4 for internal message queues between components. The message queues storing external messages had a capacity of 1 as only the head of the queue was of interest during execution (one message can be processed in an execution cycle). Regarding verification and test generation settings, we used all *model reduction* and *slicing* techniques, and *test optimization* algorithms of Gamma [14,16] to achieve the best results our framework can provide.

Each measurement was run *five* times (we calculated their *median*) on the following configuration: *Intel Core (TM) i5-1135G7 @ 2.40 GHz, DDR4 16 GB*

---

[2] Model descriptions and measurement results can be found at https://github.com/ftsrg/gamma/tree/v2.9.0/examples/, as well as [16] (RSS) and [14] (RIS).

@ *3.2 GHz, SSD 500 GB.* The model checkers were run with the following arguments:

- **Theta (DX)**: *java -jar theta –domain PRED_CART –refinement SEQ_ITP* – predicate abstraction based algorithms (most suitable for XSTS);
- **UPPAAL (XU)**: *verifyta -t0* – set to generate "some trace" (with default BFS traversal) as predictably, this is the "fastest" option (the "shortest trace" option, according to our preliminary measurements, results in "slightly" shorter generated traces but "significantly" longer generation time);
- **Spin**: we used three different sets of arguments (settings):
  1. **XP**: *spin -search -I -m250000 -w32* – DFS *approximate* iterative shortening, increased bound (larger than the state space's estimated maximum "diameter") and increased hash size for better performance: compared to the next setting, this one provides longer (not optimal, but still *sound*) traces but shorter generation time (see exact results later);
  2. **XP-i**: *spin -search -i -m250000 -w32* – *non-approximate* iterative shortening algorithm variant of the previous option;
  3. **XP-B**: *spin -search -bfs -w32* – BFS traversal of the state space.

Regarding the Spin-related arguments, the XP setting *worked* for *both models*; the other two worked only for the RSS model. For RIS, the XP-B option *ran out of memory* within 10 s, whereas XP-i *ran out* of the *300-s time limit* (timeout) for certain model-property pairs.

Table 3. Number of *test targets, generated tests* and *steps*; median end-to-end *test generation time* and *average test generation time* for a *single test target* for full *interaction* coverage in the *cascade* and *synchronous-reactive* RSS model.

	Cascade	Synchronous
#*Test targets*	49	49
#*Generated tests* (DX/XU/XP/XP-B/XP-i)	9/7/6/6/6	-/7/5/6/6
#*Steps in tests* (DX/XU/XP/XP-B/XP-i)	41/30/172/27/27	-/37/600/33/33
Σ**T** (DX/XU/XP/XP-B/XP-i) (s)	467/54/179/188/340	-/102/174/220/1102
$\overline{\text{T}}$ (DX/XU/XP/XP-B/XP-i) (s)	9.5/1.1/3.7/3.8/6.9	-/2.1/3.6/4.5/22.5

Addressing RQ-1. Table 3 shows test generation results for the RSS model aiming at covering each *interaction* [14] between the *antivalence checkers* (event raises) and *signaller* components (execution of transitions triggered by the corresponding event) in two model variants using the *cascade* and *synchronous-reactive* composition modes. As illustrated, there were 49 test targets (captured using injected boolean variables and reachability properties for each interaction [16] in the form of E F var *eventRaised* and var *correspondingTransitionFired*) that could potentially be covered by the Theta (DX), UPPAAL (XU) and Spin (XP, XP-B and XP-i) back-ends.

Regarding *test generation time*, Theta was the slowest in the cascade model variant and could not handle the more complex synchronous-reactive variant.

As for UPPAAL and Spin, the former was significantly faster (70% and 42% for XP) considering end-to-end generation time. Nonetheless, Spin seemed to scale better as the synchronous-reactive model variant did not pose a greater challenge for the XP setting compared to the cascade variant – in contrast to UPPAAL.

Regarding the *generated test size*, the XP-B and XP-i Spin settings generated the smallest test sets (least number of summed steps in tests), even though the former setting was much more efficient in terms of *generation time*. UPPAAL generated 10% longer traces on average, as expectable, considering that it uses a BFS-based model traversal mode aimed at "some trace." In addition, Theta (in the cascade variant) returned a 37% larger test set compared to UPPAAL. The XP setting returned significantly larger test sets compared to the other model checkers, even though it used the *approximate DFS iterative shortening* algorithm. In turn, these longer generated tests were utilized by the test optimization algorithms to cover the same interactions with *fewer test cases*.

To answer RQ-1, the experiment showed that even though Spin is slightly slower in terms of test generation (verification) time than its fastest counterpart (UPPAAL), it seems to be able to handle more complex interactions more efficiently, and thus, it has advantages for such models in the framework. Regarding the generated test set size, the length of the traces returned by Spin differ largely depending on the used settings. The results also show that there is a trade-off between the length of generated traces and generation time.

Addressing RQ-2. Table 4 shows test generation results aiming at covering each *state node*, *transition* and *interaction* in the asynchronous RIS model using the XU (with *array-based* mapping of message queues into UPPAAL), and the *native channel-based* (XP-N) and *array-based* (XP-A) mappings of GCL/XSTS message queues into Promela. The table does not include results for the DX mapping as Theta was unable to handle this model.

Table 4. Number of *test targets*, *generated tests* and *steps*; median end-to-end *test generation time* and *average test generation time* for a *single test target* for full *state node*, *transition* and *interaction* coverage in the integrated RIS model.

	State	Transition	Interaction
#*Test targets*	38	118	387
#*Generated tests* (XU/XP-N/XP-A)	4/5/5	26/34/34	22/24/24
#*Steps in tests* (XU/XP-N/XP-A)	30/119/119	230/870/870	240/808/808
ΣT (XU/XP-N/XP-A) (s)	243/140/135	950/1777/1653	5377/8315/7840
$\overline{\text{T}}$ (XU/XP-N/XP-A) (s)	6.4/3.7/3.6	8.1/15.1/14.0	13.9/21.4/20.3

The data show that *test generation time* increased for each verification backend as the test coverage criteria got finer. This phenomenon can be explained by the complexity of annotated model elements capturing the criteria, and the increasing number of test targets [16]. Surprisingly, Spin was on average 43% faster than UPPAAL in the case of *state node* coverage even though the result was reversed for *transition* and *interaction* coverage (45% and 34% slower on average). Regarding the two message queue representation modes, the *array-based* one was slightly faster in each case, having a 5% advantage on average.

As for the *size* of the *generated test sets*, the results were similar to that of the RSS model: Spin returned significantly longer traces compared to UPPAAL due to the underlying algorithms; however, in this case the number of generated tests was also larger. Also, there was no difference in the two message queue representation modes in this regard; Spin returned the same traces in each case.

To answer RQ-2, the experiment showed that Spin is an efficient and useful element of the framework and can provide additional flexibility in the case of different models and test coverage criteria. Regarding our message queue representation modes, our native *asynchronous channel* based solution does not bring benefits compared to our *array-based* solution. Regarding generated test set size, the results are very similar to that of RQ-1: Spin with the XP setting returns significantly longer diagnostic traces, resulting in larger test sets.

8   Conclusion and Future Work

In this paper, we presented an MBT approach for distributed (control-oriented) reactive systems with asynchronous communication. Our approach builds on the Gamma framework and features precise statechart and composition languages for component design and their integration, and hidden formal methods (model checkers) for model-based test generation. As a novelty, we integrated the open source Spin model checker to the Gamma framework via the *EXtended Symbolic Transition Systems* formalism, a verification-oriented low-level representation of reactive behavior. Our evaluation showed that Spin is efficient at generating tests and can complement the other integrated model checkers for different models.

For future work, we plan to examine the capabilities of Spin for models with parallel behavior, and how we can exploit its property language supporting linear temporal logic (LTL) [7] to capture more sophisticated test targets – two features not supported by the other integrated model checkers.

Acknowledgements. We would like to thank the anonymous reviewers for their thorough and constructive feedback. This work was partially supported by New National Excellence Program of the Ministry for Innovation and Technology, ÚNKP-23-4-I. Project no. 2019-1.3.1-KK-2019-00004 has been implemented with the support provided from the National Research, Development and Innovation Fund of Hungary, financed under the 2019-1.3.1-KK funding scheme.

Appendix: XSTS-STS Mapping

Symbolic transition system [18] models consist of two SMT [2] formulas. The *set* of *initial states* are described by a formula over the model variables, e.g., $x = 0$. The *transition relation* is described by a formula over the primed and unprimed versions of the model variables; the primed versions refer to the newly

Table 5. XSTS elements and their equivalent STS formulas.

Statement	XSTS example	STS equivalent	nextIndex
Assignment	$x := y$	x' = y	$x \to 1, y \to 0$
Assumption	**assume** $x > 5$	$x > 5$	$x \to 0$
Havoc	**havoc** x	$true$ (x' is unconstrained)	$x \to 1$
Sequence	$x := y$ **assume** $x > y$ $y := 3$	$x' = y \,\wedge$ $x' > y \,\wedge$ $y' = 3$	$x \to 1, y \to 1$
Choice	**choice** { $x := y$ } **or** { $x := y + 1$ }	$(temp = 0 \,\wedge\, x' = y) \,\vee$ $(temp = 1 \,\wedge\, x' = y + 1)$	$x \to 1, y \to 0$
If-else	**if** $(x > y)$ { $x := 0$ } **else** { $x := 1$ }	$(x > y \Rightarrow x' = 0) \,\wedge$ $(x \le y \Rightarrow x' = 1)$	$x \to 1, y \to 0$
Loop	**for** i **from** 0 **to** 1 **do** { $x := i$ }	$(i' = 0 \,\wedge\, x' = i') \,\wedge$ $(i'' = 1 \,\wedge\, x'' = i'')$	$x \to 2, i \to 2$

assigned values of the variables, e.g., $x' = x + 1$, meaning that x variable's value gets incremented by 1. To allow multiple assignments to the same variable in the transition, we allow variables to appear with more than one prime sign and annotate the transition relation with a function (*nextIndex*), describing which primed version corresponds to the variable's value in the next state after the transition; e.g., if the transition formula is $x' = 1 \wedge x'' = 2$ and the *nextIndex* function is $x \to 2$, then the x variable's value will be 2 after the transition.

Table 5 shows the mapping between the statements of the XSTS language and their equivalent STS transition formulas through examples. An *assignment* simply asserts that the next value of the left-hand-side is equal to the value of the right-hand-side. *Assumptions* translate directly into SMT formulas, while a *havoc statement* introduces the next value of the variable, but does not constrain it (therefore it can have any value from its domain).

Composite statements compose the mapping of their constituent statements. A *sequence* maps to a conjunction of its statements, but each statement will use the current primed versions of the variables. A *choice* statement requires an unconstrained temporary variable – the assignment of this variable by the solver will also determine which branch may be true (at most exactly one). An *if-else*, on the other hand, uses implication to select the branch that is asserted to be true, the else statement meaning the negation of the conjunction of all other branch conditions. A *for loop* is unfolded into a sequence, but this is done by the model checking algorithm, so the number of iterations may depend on the current state (but not the execution of the statement itself).

References

1. Ambrosi, G., Bartocci, S., Basara, L., et al.: The electronics of the high-energy particle detector on board the CSES-01 satellite. Nucl. Instrum. Methods Phys. Res. Sect. A Accel. Spectrom. Detect. Assoc. Equip. **1013**, 165639 (2021). https://doi.org/10.1016/j.nima.2021.165639
2. Barrett, C., Tinelli, C.: Satisfiability modulo theories. In: Handbook of Model Checking, pp. 305–343. Springer, Cham (2018). https://doi.org/10.1007/978-3-319-10575-8_11

3. Basu, A., et al.: Rigorous component-based system design using the BIP framework. IEEE Softw. **28**(3), 41–48 (2011). https://doi.org/10.1109/MS.2011.27
4. Behrmann, G., et al.: UPPAAL 4.0. In: Proceedings of the 3rd International Conference on the Quantitative Evaluation of Systems, QEST 2006, pp. 125–126. IEEE Computer Society, USA (2006). https://doi.org/10.1109/QEST.2006.59
5. Boulanger, J.L.: CENELEC 50128 and IEC 62279 Standards. Wiley, Hoboken (2015)
6. DeLine, R., Leino, K.R.M.: BoogiePL: a typed procedural language for checking object-oriented programs. Technical report. MSR-TR-2005-70, Microsoft Research (2005)
7. Emerson, E.A., Halpern, J.Y.: "Sometimes" and "not never" revisited: on branching versus linear time temporal logic. J. ACM **33**(1), 151–178 (1986). https://doi.org/10.1145/4904.4999
8. Enoiu, E.P., Čaušević, A., Ostrand, T.J., Weyuker, E.J., Sundmark, D., Pettersson, P.: Automated test generation using model checking: an industrial evaluation. Int. J. Softw. Tools Technol. Transfer **18**(3), 335–353 (2016). https://doi.org/10.1007/s10009-014-0355-9
9. Ferrari, A., Mazzanti, F., Basile, D., ter Beek, M.H.: Systematic evaluation and usability analysis of formal methods tools for railway signaling system design. IEEE Trans. Software Eng. **48**(11), 4675–4691 (2022). https://doi.org/10.1109/TSE.2021.3124677
10. Fraser, G., Wotawa, F., Ammann, P.E.: Testing with model checkers: a survey. Softw. Test. Verification Reliab. **19**(3), 215–261 (2009). https://doi.org/10.1002/stvr.402
11. Golarits, Z., Sinka, D., Jávor, A.: Proris - a new interlocking system for regional and moderate-traffic lines. Signal+DRAHT-Signal. Datacommun. (114), 28–36 (2022)
12. Graics, B.: Documentation of the Gamma Statechart Composition Framework v0.9. Technical report, Budapest University of Technology and Economics, Department of Measurement and Information Systems (2016). https://tinyurl.com/yeywrkd6
13. Graics, B.: Mixed-semantic composition and verification of reactive components. Technical report, Budapest University of Technology and Economics, Department of Measurement and Information Systems (2023). https://tinyurl.com/2p9dae58
14. Graics, B., Majzik, I.: Integration test generation and formal verification for distributed controllers. In: Renczes, B. (ed.) Proceedings of the 30th Ph.D. Minisymposium. Budapest Univesity of Technology and Economics, Department of Measurement and Information Systems (2023). https://doi.org/10.3311/minisy2023-001
15. Graics, B., Molnár, V.: Formal compositional semantics for Yakindu statecharts. In: Pataki, B. (ed.) Proceedings of the 24th PhD Mini-Symposium, pp. 22–25. Budapest University of Technology and Economics, Department of Measurement and Information Systems, Budapest, Hungary (2017)
16. Graics, B., Molnár, V., Majzik, I.: Integration test generation for state-based components in the Gamma framework. Preprint (2022). https://tinyurl.com/4dhubca4
17. Graics, B., Molnár, V., Vörös, A., Majzik, I., Varró, D.: Mixed-semantics composition of statecharts for the component-based design of reactive systems. Softw. Syst. Model. **19**(6), 1483–1517 (2020). https://doi.org/10.1007/s10270-020-00806-5
18. Hajdu, Á., Tóth, T., Vörös, A., Majzik, I.: A configurable CEGAR framework with interpolation-based refinements. In: Albert, E., Lanese, I. (eds.) FORTE 2016. LNCS, vol. 9688, pp. 158–174. Springer, Cham (2016). https://doi.org/10.1007/978-3-319-39570-8_11

19. Harel, D.: Statecharts: a visual formalism for complex systems. Sci. Comput. Program. **8**(3), 231–274 (1987). https://doi.org/10.1016/0167-6423(87)90035-9
20. Hartman, A., Nagin, K.: The AGEDIS tools for model based testing. ACM Sigsoft Softw. Eng. Notes **29** (2004). https://doi.org/10.1145/1007512.1007529
21. Heineman, G.T., Councill, W.T.: Component-based software engineering. Putting the Pieces Together. Addison Wesley (2001). https://doi.org/10.5555/379381
22. Henzinger, T.A., Majumdar, R.: A classification of symbolic transition systems. In: Reichel, H., Tison, S. (eds.) STACS 2000. LNCS, vol. 1770, pp. 13–34. Springer, Heidelberg (2000). https://doi.org/10.1007/3-540-46541-3_2
23. Holzmann, G.: The SPIN Model Checker: Primer and Reference Manual, 1st edn. Addison-Wesley Professional, Harlow (2011)
24. Holzmann, G.: The model checker SPIN. IEEE Trans. Software Eng. **23**(5), 279–295 (1997). https://doi.org/10.1109/32.588521
25. Huang, L.: The past, present and future of railway interlocking system. In: 2020 IEEE 5th International Conference on Intelligent Transportation Engineering (ICITE), pp. 170–174 (2020). https://doi.org/10.1109/ICITE50838.2020.9231438
26. Jéron, T., Morel, P.: Test generation derived from model-checking. In: Halbwachs, N., Peled, D. (eds.) CAV 1999. LNCS, vol. 1633, pp. 108–122. Springer, Heidelberg (1999). https://doi.org/10.1007/3-540-48683-6_12
27. Ke, X., Sierszecki, K., Angelov, C.: COMDES-II: a component-based framework for generative development of distributed real-time control systems. In: 13th IEEE International Conference on Embedded and Real-Time Computing Systems and Applications (RTCSA), pp. 199–208 (2007). https://doi.org/10.1109/RTCSA.2007.29
28. Legeard, B., Bouzy, A.: Smartesting CertifyIt: model-based testing for enterprise IT. In: 2013 IEEE Sixth International Conference on Software Testing, Verification and Validation, pp. 391–397 (2013). https://doi.org/10.1109/ICST.2013.55
29. Li, J., Post, M., Wright, T., Lee, R.: Design of attitude control systems for CubeSat-class nanosatellite. J. Control Sci. Eng. **2013** (2013). https://doi.org/10.1155/2013/657182
30. Li, W., Le Gall, F., Spaseski, N.: A survey on model-based testing tools for test case generation. In: Itsykson, V., Scedrov, A., Zakharov, V. (eds.) TMPA 2017. CCIS, vol. 779, pp. 77–89. Springer, Cham (2018). https://doi.org/10.1007/978-3-319-71734-0_7
31. Lukács, G., Bartha, T.: Formal modeling and verification of the functionality of electronic urban railway control systems through a case study. Urban Rail Transit. **8** (2022). https://doi.org/10.1007/s40864-022-00177-8
32. Martinez, S., Pereira, D.I.D.A., Bon, P., Collart-Dutilleul, S., Perin, M.: Towards safe and secure computer based railway interlocking systems. Int. J. Transp. Dev. Integr. **4**(3), 218–229 (2020)
33. Mohalik, S., Gadkari, A.A., Yeolekar, A., Shashidhar, K., Ramesh, S.: Automatic test case generation from Simulink/Stateflow models using model checking. Softw. Test. Verif. Reliab. **24**, 155–180 (2014). https://doi.org/10.1002/stvr.1489
34. Molnár, V., Graics, B., Vörös, A., Majzik, I., Varró, D.: The Gamma Statechart Composition Framework. In: 40th International Conference on Software Engineering (ICSE), Gothenburg, Sweden, pp. 113–116. ACM (2018). https://doi.org/10.1145/3183440.3183489
35. Mondok, M.: Efficient abstraction-based model checking using domain-specific information. Technical report, Budapest University of Technology and Economics, Department of Measurement and Information Systems (2021). https://tinyurl.com/yh4b8w98

36. Owre, S., Rushby, J.M., Shankar, N.: PVS: a prototype verification system. In: Kapur, D. (ed.) CADE 1992. LNCS, vol. 607, pp. 748–752. Springer, Heidelberg (1992). https://doi.org/10.1007/3-540-55602-8_217

37. Radnai, B.: Integration of SCXML state machines to the Gamma framework. Technical report, Budapest University of Technology and Economics, Department of Measurement and Information Systems (2022). https://tinyurl.com/4mmtsw7v

38. Jéron, T., Morel, P.: Test generation derived from model-checking. In: Halbwachs, N., Peled, D. (eds.) CAV 1999. LNCS, vol. 1633, pp. 108–122. Springer, Heidelberg (1999). https://doi.org/10.1007/3-540-48683-6_12

39. Sztipanovits, J., Bapty, T., Neema, S., Howard, L., Jackson, E.: OpenMETA: a model- and component-based design tool chain for cyber-physical systems. In: Bensalem, S., Lakhneck, Y., Legay, A. (eds.) ETAPS 2014. LNCS, vol. 8415, pp. 235–248. Springer, Heidelberg (2014). https://doi.org/10.1007/978-3-642-54848-2_16

40. Tóth, T., Hajdu, A., Vörös, A., Micskei, Z., Majzik, I.: Theta: a framework for abstraction refinement-based model checking. In: Stewart, D., Weissenbacher, G. (eds.) Proceedings of the 17th Conference on Formal Methods in Computer-Aided Design, pp. 176–179 (2017). https://doi.org/10.23919/FMCAD.2017.8102257

41. Zhou, J., Hu, Q., Friswell, M.I.: Decentralized finite time attitude synchronization control of satellite formation flying. J. Guid. Control. Dyn. **36**(1), 185–195 (2013). https://doi.org/10.2514/1.56740

A Mechanized Semantics
for Component-Based Systems
in the HAMR AADL Runtime

Stefan Hallerstede[1] and John Hatcliff[2](✉)

[1] Aarhus University, Aarhus, Denmark
[2] Kansas State University, Manhattan, KS 66506, USA
hatcliff@ksu.edu

Abstract. Many visions for model-driven component-based development emphasize models as the "single source of truth" by which different forms of analysis, specification, verification, and code generation are integrated. Such a vision depends strongly on a clear modeling language semantics that provides different tools and stakeholders with a common understanding of a model's meaning. In this paper, we report on a mechanization of a formal semantics in the Isabelle theorem prover for key aspects of the SAE standard AADL modeling language. A primary goal of this semantics is to support component-oriented contract specification and verification as well as code generation implemented in the HAMR AADL model-driven development tool chain. We provide formal definitions of run-time system state, execution steps, reachable states, and property verification. Use of the mechanization for real-world applications is supported by automated HAMR translation from AADL models into the Isabelle specifications. In addition to general verification support, we define well-formedness properties and associated proofs for models, system states, and traces that are automatically proven for HAMR-generated Isabelle models.

1 Introduction

Model-driven development tools continue to gain traction for building and assuring safety-critical systems. The Architecture Analysis and Design Language (AADL) [1] stands out among other standardized modeling languages due to its stronger semantic interpretation of modeling elements. Stronger emphasis on semantics has led to AADL's use in numerous large-scale industrial research projects, particularly those prioritizing formal methods. An ecosystem of tooling has developed around AADL that includes analyses for behavior specification and verification, resource utilization and timing, trade space exploration, hazard analysis, code generation, and assurance case development.

Despite this emphasis on semantics, semantic descriptions in the current version of the standard are presently mostly in narrative form, or with limited use of timed automata that are not strongly integrated with other descriptions. This makes it more difficult to establish the soundness of model-based analyses and verification, the correctness of code generation, and a consistent semantic

J. Cámara and S.-S. Jongmans (Eds.): FACS 2023, LNCS 14485, pp. 45–64, 2024.
https://doi.org/10.1007/978-3-031-52183-6_3

interpretation across multiple tools. Recently, Hugues has led an effort within the AADL committee to develop a road map for integrating formal definitions of semantics into the standard. This is supported in part by an open source mechanization of the AADL static semantics and various supporting analyses in the Coq theorem prover [22].

In an accompanying coordinated effort, Hatcliff, Hugues and others [19] developed a rule-based specification of important elements of the AADL Run-Time Services (RTS). The AADL RTS aim to (a) hide the details of specific RTOS and communication substrates (e.g., middleware) and (b) provide AADL-aligned system implementations with canonical platform-independent actions that realize key steps in the integration and coordination of component application logic. Though it is designed to be implementation-independent, traceability to the RTS formalization in [19] has been emphasized in the High Assurance Modeling and Rapid engineering framework (HAMR) AADL code generation framework [17] as a guide in developing consistent code generation on multiple platforms and in designing integrated model and code behavior contract specification and verification [21] and property-based testing [18]. HAMR provides code generation in C and in Slang [26] (a safety-critical subset of Scala, that can also be translated to C) and system deployments can be generated for the JVM, Javascript, Linux, and the seL4 microkernel.

To improve the utility, feature coverage, and confidence in the formalization in [19], we are developing a suite of interconnected artifacts including an executable version of the semantics (to serve as an abstract reference implementation, e.g., to use in model-based testing) and an encoding of the semantics in the Isabelle theorem prover. One of HAMR's code generation targets is the seL4 micro-kernel whose semantics and correctness properties have been formally specified and proved in Isabelle – providing one of the most significant applications of formal methods to date [23]. One of the motivations for chosing Isabelle over another theorem prover is to provide semantic specifications for AADL and HAMR that can eventually be connected to the seL4 formalization.

In this paper, we report on the Isabelle-based mechanization of AADL semantics, inspired by the definitions in [19], as well as several significant novel extensions that provide foundations for verification and refinements of the framework to address characteristics of HAMR code generation target platforms. We refer to this mechanization as AADL-HSM (AADL HAMR Semantics Mechanization). The specific contributions of this paper are as follows.

- We provide a mechanization of the notions of port, thread, and system states as well as the AADL run-time services given in [19].
- We give definitions of AADL communication that accommodate AADL port-based communication properties and that can be instantiated to the communication semantics of different deployment platforms.
- We give definitions for thread scheduling that can accomodate, e.g. the static cyclic scheduling used by HAMR on recent industry projects [5].
- We introduce basic definitions of a property framework for reasoning about threads and system properties. This lays the formal foundation for addressing

the rationale and soundness for the GUMBO contract framework for component verification [21] and testing [18].
- We extend the HAMR code generator to translate core AADL instance models and HAMR initial system states into the mechanization.

This paper summarizes the approach and key aspects of the above contributions, but the primary artifacts substantiating the results are our Isabelle specifications publicly available at [29]. These are formatted and commented using Isabelle's documentation framework to provide a 100+ page PDF readable guide to the semantics of the AADL subset supported by HAMR. The HAMR distribution is available at [28].

2   AADL Background

SAE International standard AS5506C [1] defines the AADL core language for expressing the structure of embedded, real-time systems via definitions of components, their interfaces, and their communication.

AADL provides a precise, tool-independent, and standardized modeling vocabulary of common embedded software and hardware elements using a component-based approach. Components have a category that defines a standard interpretation. Categories include software (e.g., threads, processes), hardware (e.g., processor, bus), and system (interacting hardware and software). Each category also has a distinct set of standardized properties that can be used to configure the specific component's semantics for various aspects: timing, resources, etc.

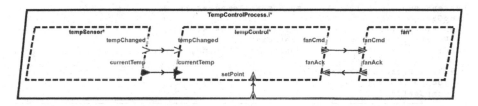

Fig. 1. Temperature Control Example (excerpts) – AADL Graphical View

Figure 1 presents a portion of the AADL standard graphical view for a simple thermostat that maintains a temperature according to a set point structure. The system (not shown) contains a process called tempControlProcess. This process consists of three threads: tempSensor, tempControl, and fan, shown in the figure from left to right. Passing data from one thread to another is done through the use of ports. Each port can be classified as an *event port* (e.g., to model interrupt signals or other notification-oriented messages without payloads), a *data port* (e.g. modeling shared memory between components or distributed memory services where an update to a distributed memory cell is automatically propagated to other components that declare access to the cell), or an *event data port* (e.g., to model asynchronous messages with payloads, such as in publish-subscribe frameworks). We refer to these classifications as the port's *kind*. Inputs to event and event data ports are buffered. The buffer sizes and overflow policies can be configured per port using standardized AADL properties. Because of these similarities, we introduce the term *event-like* to refer collectively to event and event data ports. Inputs to data ports are not buffered; newly arriving data overwrites the previous value.

The periodic `tempSensor` thread measures the current temperature, e.g., from memory-mapped IO, which is not shown in the diagram, and transmits the reading on its `currentTemp` data port. If it detects that the temperature has changed since the last reading, then it sends a notification to the sporadic (that is, *event driven*) `tempControl` thread on its `tempChanged` event port. When the `tempControl` thread receives a `tempChanged` event, it will read the value on its `currentTemp` data port and compare it with the most recent set points. If the current temperature exceeds the high set point, it sends a command to the `fan` thread to turn `on`. Similarly, if the current temperature is low, it will send an `off` fan command. In either case, `fan` acknowledges whether it was able to fulfill the command by sending user-defined data types `FanAck.Ok` or `FanAck.Error` on its `fanAck` event data port.

AADL provides a textual view to accompany the graphical view. The following listing illustrates the *component type* declaration for the `tempControl` thread for the example above. The data transmitted over ports are of specific types, and properties such as `Dispatch_Protocol` and `Period` configure the tasking semantics of the thread.

```
thread TempControl
features
 currentTemp: in data port TempSensor::Temperature.i;
 tempChanged: in event port;
 fanAck: in event data port CoolingFan::FanAck;
 setPoint: in event data port SetPoint.i;
 fanCmd: out event data port CoolingFan::FanCmd;
properties
 Dispatch_Protocol => Sporadic;
 Period => 500 ms;   -- the min sep between incoming msgs
end TempControl;
```

The next listing illustrates integration of subcomponents. The body of process `TempControlProcess` type has no declared features because the component does not interact with its context in this simplified example. The implementation of `TempControlProcess` specifies subcomponents and subcomponent communications over declared connections between ports.

```
process TempControlProcess
 -- no features; no interaction with context
end TempControlProcess;

process implementation TempControlProcess.i
subcomponents
 tempSensor : thread TempSensor::TempSensor.i;
 fan : thread CoolingFan::Fan.i;
 tempControl: thread TempControl.i;
 operatorInterface: thread OperatorInterface.i;
```

```
connections
c1:port tempSensor.currentTemp -> tempControl.currentTemp;
c2:port tempSensor.tempChanged -> tempControl.tempChanged;
c3:port tempControl.fanCmd -> fan.fanCmd;
c4:port fan.fanAck -> tempControl.fanAck;
end TempControlProcess.i;
```

AADL provides many standard property sets including those used to configure the core semantics of AADL, e.g., thread dispatching or port-based communication. Developer-specified property sets enable one to define project-specific configuration parameters.

In this paper (following [19]) and in our AADL-HSM mechanization in Isabelle, we limit our semantics presentation to thread components since almost all of AADL's run-time services are associated with threads and port-based communication.[1]

3 Model Representation

The HAMR tool chain uses a similar approach to many AADL tools: it processes an AADL *instance model* [15, Section 4.1.6] as generated by the AADL OSATE Integrated Development Environment (IDE). The instance model indicates specific component implementations to be used in a system build, and it provides connection topologies directly in terms of thread component ports (thus "flattening" the model and emphasizing threads as the primary run-time entities of the system). Within its generated code, HAMR includes data structures representing model information such as thread interfaces, port declarations, connection information, and associated AADL properties referenced within the run-time system. AADL-HSM has a corresponding representation of model information centered around thread and port descriptors, *CompDescr* and *PortDescr* respectively,

record *CompDescr* =
   name :: *string*
   id :: *CompId*
   portIds :: *PortIds*
   dispatchProtocol :: *DispatchProtocol*
   dispatchTriggers :: *PortIds*
   compVars :: *Vars*

record *PortDescr* =
   name :: *string*
   id :: *PortId*
   compId :: *CompId*
   direction :: *PortDirection*
   kind :: *PortKind*
   size :: *nat*
   urgency :: *nat*

Like many modeling tools, HAMR autogenerates unique identifiers for model elements. Types are introduced for component identifiers and port identifiers in the code, and corresponding types *CompId* and *PortId* appear in the AADL-HSM definitions. *PortIds* is a type for a set of port identifiers. The top-level *Model* structure contains a mapping from component identifiers to component descriptors (similarly for ports),

record *Model* =
   modelCompDescrs :: (*CompId, CompDescr*) *map*
   modelPortDescrs :: (*PortId, PortDescr*) *map*
   modelConns :: *Conns*

---

[1] Other categories of components, see [1], will be treated in later work.

Connections are represented as a map from a connection source *PortId* to a set of one or more target *PortIds*,

type-synonym *Conns* = (*PortId*, *PortIds*) *map*

Functions *isInCDPID* and *isInCIDPID* are two of approximately 40 helper functions for accessing model elements in AADL-HSM specifications:

fun *isInCDPID* :: *Model* ⇒ *CompDescr* ⇒ *PortId* ⇒ *bool*
    where *isInCDPID* *m* *cd* *p* = (*p* ∈ *portIds* *cd* ∧ *isInPD*(*modelPortDescrs* *m* $ *p*))
fun *isInCIDPID* :: *Model* ⇒ *CompId* ⇒ *PortId* ⇒ *bool*
    where *isInCIDPID* *m* *c* = *isInCDPID* *m* (*modelCompDescrs* *m* $ *c*)

These two examples are predicates that hold for a model *m* when a port with identifier *p* is an input port in a component with descriptor *cd* (or alternatively for a component with identifier *c*). The functions use Isabelle record, set, and map operations. Isabelle maps are defined as functions from a "*key*" domain to an option "*val*" domain of type "*key* ⇒ *option val*". We add an infix operator $ whose application will reduce (via the Isabelle simplifier tactic) to a value *v* when a map lookup operation returns *Some v* (if lookup returns *None*, then the expression will fail to produce a canonical value).

A number of model well-formedness properties are included with the base model specification. The following example property specifies that connections only go from *out* ports to *in* ports of matching kinds:

definition *wf-Model-ConnsPortCategories* :: *Model* ⇒ *bool*
    where *wf-Model-ConnsPortCategories* *m* ≡
    (∀ *p* ∈ *dom* (*modelConns* *m*). *isOutPID* *m* *p* ∧
    (∀ *p'* ∈ (*modelConns* *m* $ *p*). *isInPID* *m* *p'* ∧ (*kindPID* *m* *p* = *kindPID* *m* *p'*)))

When HAMR translates a system representation into AADL-HSM, it also generates proofs that the model satisfies the well-formedness properties (which are first established by the HAMR OSATE IDE plug-in). Due to the architecture of the theory, these proofs only need to use the Isabelle simplifier tactics. This organization provides the foundation for specializations of the semantics to be defined (e.g., for AADL subsets that only use a particular set of features, or that require AADL properties relevant to a particular development pipeline). For example, by adding additional well-formedness properties, we can indicate the modeling sublanguage used on the DARPA CASE program for an seL4-based platform. [29] illustrates model-based specifications and well-formedness proofs for several system examples.

4   State Representation and Application Logic

A key contribution of [19] was a mathematical description of key elements of an AADL-based system's run-time state. In this section, we summarize our mechanization of those notions as well as various enhancements and extensions. Figure 2 presents the graphical summary from [19] of thread state concepts. Many of AADL's thread execution concepts are based on long-established task patterns and principles for achieving analyzeable real-time systems [9]. Following

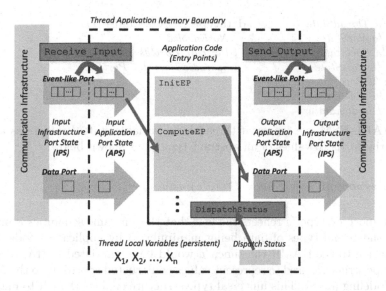

Fig. 2. Thread and Port State Concepts

these principles, at each activation of a thread, the application code of the thread will abstractly compute a function from its input port values and local variables to output port values while possibly updating its local variables. In AADL terminology, *dispatching* a thread refers to the thread becoming ready for execution from a OS scheduler perspective, and the semantics for dispatch is determined by several thread properties specified in the AADL (captured formally as indicated in the definition *CompDescr* above). The thread `Dispatch_Protocol` property selects among several strategies for determining when a thread should be dispatched. In this paper, we consider only `Periodic`, which dispatches a thread when a certain time interval is passed, and `Sporadic`, which dispatches a thread upon arrival of messages to input ports specified as *dispatch triggers*. When a thread is dispatched, information describing the reason for its dispatch is stored in the thread's state and is retrievable via the `Dispatch_Status` RTS (diagrammed in Fig. 2 and formalized below). For example, in a sporadic component, `Dispatch_Status` returns information indicating which port triggered the dispatch. This may be used in either the component application or infrastructure code to branch to a message handler method dedicated to processing messages arriving on the particular port.

Figure 2 illustrates that a thread's state includes the state of its ports, local variables, and dispatch status, and this is formalized in the *ThreadState* definition below.

record $'a$ *ThreadState* =
  tvar :: $'a$ *VarState*
  infi :: $'a$ *PortState*
  appi :: $'a$ *PortState*
  appo :: $'a$ *PortState*
  info :: $'a$ *PortState*
  disp :: *DispatchStatus*

datatype *DispatchStatus* =
  NotEnabled
  | *Periodic PortIds*
  | *Sporadic PortId* * *PortIds*

In AADL-HSM, we represent thread local variable state *VarState* as a map from variable ids to values of an abstract type $'a$ as shown below.

type-synonym $'a$ *VarState* = (*Var*, $'a$) *map*

The polymorphic type $'a$ reflects the fact that the run-time semantics is orthogonal to the actual types of data being manipulated by application code. In an instantiation to the HAMR run-time, $'a$ would be instantiated to HAMR's universal type structure used to represent values constructed according to the AADL Data Modeling annex. This universal type structure enables HAMR to map values to and provide interoperability between the different programming language type systems supported by HAMR code generation.

One of the most important aspects of the thread state are the various buffers and memory blocks used to storing incoming and outgoing data for ports. In AADL-HSM, these are represented uniformally using values from the *PortState* type, which map from *PortIds* to $'a$ *Queues*.

type-synonym $'a$ *PortState* = (*PortId*, $'a$ *Queue*) *map*

Queues encapsulate lists with a bound and overflow policy as described in the AADL standard. We enhance the definitions in [19] with various properties proven about queue behavior.

Figure 2 illustrates that the state of each port is decomposed into the Infrastructure Port State (IPS) and the Application Port State (APS). The IPS represents the communication infrastructure's perspective of the port. The APS represents the thread application code's perspective of the port. The *Thread-State* definition includes input IPS and APS (*infi* and *appi*) and output IPS and APS (*info* and *appo*). The distinction between IPS and APS is used to represent AADL's notions of *port freezing* and *port variable* as presented in more detail in [14].

Typically, when a thread is dispatched, the component infrastructure uses the `Receive_Input` RTS (referenced in Sect. 5) to move one or more values from the IPS of input ports into the input APS. Then the component application code is called and the APS values remain "frozen" as the code executes. This provides the application a consistent view of inputs even though input IPS may be concurrently updated by communication infrastructure behind the scenes. The application code writes to the output APS throughout execution. Our intended design for this is that when the application code completes, the component infrastructure will call the `Send_Output` RTS to move output values from the output APS to the IPS, thus releasing the output values all at once to the communication infrastructure for propagation to consumers.

For input event data ports, the IPS typically would be a queue into which the middleware would insert arriving values following overflow policies specified for the port. For input data ports, the IPS typically would be a memory block large enough to hold a single value. For output ports, the IPS represents pending value(s) to be propagated by the communication infrastructure to connected consumer ports. At the component's external interface, this execution pattern follows the *Read Inputs; Compute; Write Outputs* structure championed by various real-time system methods (e.g., [9]) enabling analyzeability.

The AADL standard indicates that a thread's application code is organized into entry points (e.g., subprograms that are invoked from the AADL run-time). For example, the *Initialize Entry Point* (`InitEP`) is called during the system's initialization phase, the *Compute Entry Point* (`ComputeEP`) is called during the system's "normal" compute phase. The behavior of entry point application logic is formalized in an *App* record as constraints on the thread state variables and the application port state.

record *'a App* =
   appInit :: *'a VarState* ⇒ *'a PortState* ⇒ *bool*
   appCompute :: *'a VarState* ⇒ *'a PortState* ⇒ *DispatchStatus* ⇒
       'a VarState ⇒ *'a PortState* ⇒ *bool*

The definitions indicate that a thread's `ComputeEP` behavior relates its input application port state, variable state, and dispatch status, to output application port state and possible updated variable state. The Initialize entry point has no "inputs"; it simply provides initial values for the output application port state and variable state. These definitions are strongly aligned with the GUMBO contract language for AADL. [21] illustrates how the Logika SMT-based verifier can automatically verify that a thread's component entry point code conforms to thread contracts. Alternatively, property-based testing can test that implementations conform to executable versions of the contracts [18]. Semantically, the contracts for each thread collectively give rise to constraints on the thread state as typed above. In a preliminary investigation, we have hand-translated GUMBO contracts into Isabelle using a shallow embedding. A fully automated translation of contracts will eventually enable us to carry out Isabelle proofs of system properties based on component behaviors derived from contracts that have been tested or proven, e.g., by Logika, to correctly summarize component behavior code.

Each thread is associated to its application logic behavior via a mapping from *CompIds* to *Apps*.

type-synonym *'a CIDApp* = (*CompId, 'a App*) *map*

As one example of a number of well-formed state properties, threads must only access their thread-local variable and port states as specified in the *Model*. This can be exspressed for a *CIDApp ca*, a *Model m*, a thread component *c* of the model *m* and an application *a* with *a* = *ca* $ *c*,

$\forall$ vs ps d ws qs. appCompute a vs ps d ws qs $\longrightarrow$
  ($\forall$ v. v $\in$ dom vs $\cup$ dom ws $\longrightarrow$ isVarOfCID m c v)

We wish to distinguish the specifications that the developer supplies for an application (e.g., model information and thread application logic) from the state and execution logic of the AADL run-time and underlying execution platform. Accordingly, we introduce a record type *AppModel* for developer-supplied information that aggregates the structural model information (*Model*) and behavior specifications for each thread (*CIDApp*). Further separating the developer's structural specifications and behavior specifications permits AADL-HSM to support an approach for model refinement where the structural model remains unchanged but the application logic can be considered at different abstraction levels.

record '*a AppModel* =
  appModel :: *Model*
  appModelApp :: '*a CIDApp*

Finally, the system state includes thread states (*systemThread*), the state of the communication substrate infrastructure (*systemComms*), and information for phasing and scheduling as discussed in Sect. 5.

record ('*u*, '*a*) *SystemState* =
  systemThread :: (*CompId*, '*a ThreadState*) *map*
  systemComms :: '*u*
  systemState :: (*CompId*, *ScheduleState*) *map*
  systemPhase :: *Phase*
  systemExec :: *Exec*

datatype *Phase* =
  Initializing | *Computing*
datatype *ScheduleState* =
  Waiting | *Ready* | *Running*
datatype *Exec* =
  Initialize CompId list |
  Compute CompId

systemComms is designed as an abstraction that can be instantiated to specific behavior rules based on the nature of a particular deployment platform (e.g., distributed communication via middleware, local communication via shared memory) and even different fault models regarding message loss, reordering, etc. This also supports the AADL standard's philosophy of not specifying precisely the implementation of communication, but instead constraining it (or stating assumptions about it), using various model properties. We only require particular instantiations to provide operations *comPush* and *comPull* to push data into the substrate and pull data out of the substrate. The two operations share the same signature. Taking the substrate state '*u*, a port state of a component '*a Portstate* and the port connections *Conns*, they yield a set of possible updates to the substrate and the port state. The signatures are symmetric and pass all available information to achieve a high degree of freedom for different instantiations.

record ('*u*, '*a*) *Communication* =
  comPush :: '*u* $\Rightarrow$ '*a PortState* $\Rightarrow$ *Conns* $\Rightarrow$ ('*u* $\times$ '*a PortState*) *set*
  comPull :: '*u* $\Rightarrow$ '*a PortState* $\Rightarrow$ *Conns* $\Rightarrow$ ('*u* $\times$ '*a PortState*) *set*

The least constraining substrate *CommonComm* permits any amount of pushing into and pulling out of the substrate (including doing nothing) that respect the queue capacities of the ports. The representation follows that of [22], storing tuples $(p,\ v,\ p',\ t)$ where p is the sending port, v is a value, p' is the receiving port and t is a token that gives a unique identity to each message in the substrate. *CommonComm* does not impose any ordering of messages in the substrate.

definition *CommonComm* :: $((PortId * 'a * PortId * nat)\ set,\ 'a)$ *Communication*
  where *CommonComm* = $($ *comPush=* ..., *comPull=* ... $)$

The motivation for this approach to expressing the substrate is to support verification of very basic properties using *CommonComm* and require other substrates to refine it. In particular, well-formedness properties of the state should be inherited from *CommonComm* so that they do not have to be proved for other more specific substrates related to specific AADL execution and communication models (e.g., AADL's distinction between delayed and immediate communication).

5   System Behavior and Properties

As noted in [19], the current AADL standard underspecifies coordination between threads and the underlying scheduling and communications. The standard uses hybrid automata to specify constraints and timing aspects on the operational life-cycle of a thread (e.g., through initialization, compute, and finalization phases, along with mode changes and error recovery). Guarded transitions in the automata correspond to checks on the thread state, interactions with the scheduler, etc. Since the focus of the automaton is on a single thread, broader aspects of the system state, including the scheduling dimension and communication substrate, are not reflected in the standard.

In addition to specifying how HAMR interprets these concepts from the standard, AADL-HSM embodies a proposal to the broader AADL community for how these concepts may be formalized. Reflecting the standard's emphasis on the thread operational life-cycle, AADL-HSM first presents rules[2] corresponding to transitions in the standard's automata.[3] One important contribution of our work that supports reasoning about component and system behaviors is a formalization of application logic (which was not addressed in the standard) in the definition of a thread's execution steps.

Thread Behavior: The following rules illustrate some of the key aspects of thread execution. The AADL standard specifies that system execution is organized into (a) an *initialization* phase during which thread Initialize entry points are executed to initialize thread local variables and output ports and (b) a *compute* phase in which thread *Compute* entry points are executed according to

---

[2] The rules are defined inductive to have access to the associated proof support that Isabelle offers. Sequences of transitions are modeled by transitive closure of the rules.

[3] Our life-cycle steps are simplified because we omit AADL mode switching and error recovery since HAMR does not currently support these features.

the scheduling policy. We divide the thread execution step rules to match this phasing.

The main rule for thread execution in the initialization phase shown below "lifts" the thread application logic (which only constrains the thread local variables *tvar* and application's view of output ports *appo*) to the entire thread state.

inductive *stepInit* **for** $a :: 'a\ App$ **and** $t :: 'a\ ThreadState$
    where *initialize*: $applnit\ a\ vs\ ps \implies stepInit\ a\ t\ (t(\!|\ tvar:=\ vs,\ appo:=\ ps\ |\!))$

The main rule for thread execution in the compute phase is shown below. The rule is parameterized on several auxiliary elements. t is the thread state for the thread undergoing a transition. Given the input infrastructure port state ($infi$) for the thread, the function cd computes a set of dispatch status for the thread – indicating the possible scenarios in which it is dispatchable. We realize cd as a collection of Isabelle functions that formalizes the informally described rules in the AADL standard for thread dispatch. ap is the application logic for the thread. ca is a relation indicating that transitions between the thread's scheduling state (roughly corresponding to states in the life-cycle diagrams presented in the AADL standard). The three rules *dispatch*, *compute*, and *complete* are constrained to follow the life cycle evolution order. *dispatch* moves an enabled thread from its "waiting for dispatch" state to be ready for execution. The *receiveInput* auxillary rule formalizes the AADL RTS Receive Input as shown in Fig. 2 used to move incoming data from the infrastructure ports of thread ($infi$) into the view of the application logic (in $appi$) The specific ports to receive data on are retrieved from the dispatch status via the *dispatchInputPorts* function (see [19] for a detailed discussion). The *compute* rule represents the execution step of thread task, and relationship between application input ports, variables, dispatch status, and output ports is determined by the application logic. In the post state, the input application port state is cleared. The *complete* rule releases the output port contents to the infrastructure, and moves the thread back to the *waiting* state. Together, the rules integrate the Receive Input RTS, the application logic, and Send Output RTS, to realize the Read Inputs; Compute; Write Outputs pattern described in Sect. 4.

inductive *stepThread* **for** $cd :: 'a\ PortState \Rightarrow DispatchStatus\ set$
                and $pk :: PortId \Rightarrow PortKind$
                and $ap :: 'a\ App$
                and $ca :: ScheduleState * ScheduleState$
                and $t :: 'a\ ThreadState$ **where**
dispatch: $[\![\ ca = (Waiting,\ Ready);\ dsp \in cd\ (infi\ t);\ dsp \neq NotEnabled;$
        $receiveInput\ pk\ (dispatchInputPorts\ dsp)\ (infi\ t)\ (appi\ t)\ infi'\ appi'\ ]\!]$
  $\implies stepThread\ cd\ pk\ ap\ ca\ t\ (t(\!|\ infi:=\ infi',\ appi:=\ appi',\ disp:=\ dsp\ |\!))\ |$
compute: $[\![\ ca = (Ready,\ Running);\ appCompute\ ap\ (tvar\ t)\ (appi\ t)\ (disp\ t)\ ws\ qs\ ]\!]$
  $\implies stepThread\ cd\ pk\ ap\ ca\ t\ (t(\!|\ tvar:=\ ws,\ appo:=\ qs,$
                            $appi:=\ clearAll\ (dom\ (appi\ t))\ (appi\ t)\ |\!))\ |$
complete: $[\![\ ca = (Running,\ Waiting);\ sendOutput\ (appo\ t)\ (info\ t)\ appo'\ info'\ ]\!]$
  $\implies stepThread\ cd\ pk\ ap\ ca\ t\ (t(\!|\ appo:=\ appo',\ info:=\ info',$
                      $disp:=\ NotEnabled\ |\!))$

System Behavior: The following rules illustrate some of the key aspects of system execution. This presentation is a significant evolution in design beyond the non-mechanized rule excepts in [19]. The rules are parameterized on the combined model structure and application logic, the semantics for the communication substrate, and the system state. To connect with how AADL / HAMR was applied on the recent DARPA CASE project with Collins Aerospace, we show a variant of the rules instantiated to a static cyclic scheduling regime where *sc* is a data structure specifying the schedule.

inductive *stepSys* **for** *am* :: *'a AppModel*
    and *cm* :: *('u, 'a) Communication*
    and *sc* :: *SystemSchedule*
    and *s* :: *('u, 'a) SystemState* **where**
initialize: ⟦ *isInitializing s*; *systemExec s* = *Initialize* (*c#cs*);
    stepInit (*appModelApp am* $ *c*) (*systemThread s* $ *c*) *t* ⟧
    ⟹ *stepSys am cm sc s* (*s*⦇ *systemThread*:= (*systemThread s*)(*c*↦*t*),
                systemExec:= *Initialize cs* ⦈)) |
switch: ⟦ *isInitializing s*; *systemExec s* = *Initialize* []; *c* ∈ *scheduleFirst sc* ⟧
    ⟹ *stepSys am cm sc s* (*s*⦇ *systemPhase*:= *Computing*,
                systemExec:= *Compute c* ⦈)) |
push: ⟦ *isComputing s*; *systemThread s c* = *Some t*;
    (*sb*, *it*) ∈ *comPush cm* (*systemComms s*) (*info t*) (*appModelConns am*) ⟧
    ⟹ *stepSys am cm sc s* (*s*⦇ *systemComms*:= *sb*,
                systemThread:= (*systemThread s*)(*c*↦(*t*⦇ *info*:= *it* ⦈)))⦈)) |
pull: ⟦ *isComputing s*; *systemThread s c* = *Some t*;
    (*sb*, *it*) ∈ *comPull cm* (*systemComms s*) (*infi t*) (*appModelConns am*) ⟧
    ⟹ *stepSys am cm sc s* (*s*⦇ *systemComms*:= *sb*,
                ꜱyꜱꜰeₘThꞟ ⱺuⱡ.⩵ (ꜱyꜱꜰeₘThꞟ ⱺuⱡ ꜱ)(ⱺꞟ ⸱(ⱡ⦇ ⁱⁿfⱡ.⩵ ⁱⱡ ⦈)))⦈)) |
execute: ⟦ *isComputing s*; *systemExec s* = *Compute c*; *c'* ∈ *scheduleComp sc* $ *c*;
    stepThread (*computeDispatchStatus* (*appModel am*) *c*)
        (*appModelPortKind am*) (*appModelApp am* $ *c*)
        (*systemState s* $ *c, a*) (*systemThread s* $ *c*) *t* ⟧
    ⟹ *stepSys am cm sc s* (*s*⦇ *systemThread*:= (*systemThread s*)(*c*↦*t*),
                systemState:= (*systemState s*)(*c*↦*a*),
                systemExec:= *Compute c'* ⦈))

The *initialize* rule can apply when the system is in the initialization phase (*isInitializing s*): executing the Initialize Entry Point code of a thread component with id *c*, is represented by applying the application logic thread using the previously defined *stepInit* rule. Given that the Initialize entry point code does not read any state values, we are able to formally prove that the ordering of threads given by the thread id initialization list *Initialize cs* is irrelevant. The *switch* rule moves the system from initialization phase to the compute phase when the thread initialization list is empty.

Within the compute phase (*isComputing s*), in this most general (i.e., most abstract) version of the semantics, we do not constrain communication steps to a particular system schedule, which corresponds to the expected behavior, e.g., when using an independent middleware framework like the OMG Data Distribution Service (DDS) with its own notion of threading to move data between components. Thus, rules for moving data onto the communication substrate from a thread's output infrastructure ports (*push*) and for moving data off of the communication substrate to a thread's input infrastructure ports (*push*) are allowed to interleave arbitrarily with the execution steps of a thread (*execute*). In the *execute* rule, the thread with id c that appears next in the static schedule is stepped using the *stepThread* rule. The system state is updated to reflect the updated thread state, system state, and next thread component to be scheduled.

As indicated in Sect. 4, these rules are designed to be abstract, with various properties being provable about the most general case, but then refined to more specific notions of execution where stronger properties can be proved. For example, the most general communication rules do not guarantee message delivery, or any notion of delivery fairness.

Given the rules above, notions of transitive system transitions and reachable states can be defined as follows.

definition *stepsSys* **where** *stepsSys am cm sc* = (*stepSys am cm sc*)**
definition *reachSys* **where**
   reachSys am cm sc y $\equiv$ $\exists x.\ initSys\ (appModel\ am)\ x \wedge stepsSys\ am\ cm\ sc\ x\ y$

Property Framework: Following the semantics definitions above, properties of AADL models can be verified starting provided definitions of application logic (e.g., as might be automatically derived from AADL GUMBO contracts for each thread component). For instance, each application might establish some property P as described by *appInitProp*.

definition *appInitProp* **where**
   appInitProp a P $\equiv \forall s'\ p'.\ appInit\ a\ s'\ p' \longrightarrow P\ (s',\ p')$

This implies that a family of such properties $P\ c$ must be established for all thread components c on system level.

definition *sysInitProp* **where** *sysInitProp am P* $\equiv \forall c \in appModelCIDs\ am.$
   $\forall t\ t'.\ stepInit\ (appModelApp\ am\ \$\ c)\ t\ t' \longrightarrow P\ c\ (tvar\ t',\ appo\ t')$

A lemma of the accompanying theory characterises this relationship between application initialisations more precisely. Given a well-formed AADL instance model, a family of system initialisation properties is established, if we can prove that each initialisation establishes one of those. The latter is described as a family of verification conditions that must be established for the property on the system level to hold.

lemma *initSysFromApps*:
  assumes *wf*: *wf-AppModel am*
  and *vc*: $\bigwedge c.\ c \in appModelCIDs\ am \implies appInitProp\ (appModelApp\ am\ \$\ c)\ (P\ c)$
  shows *sysInitProp am P*

This approach permits to reason as much as possible locally on the level of applications lifting these properties to the system level, ignoring its complexity at this stage. One can proceed similarly with the computation parts of applications, starting with application properties, and so on.

definition *appInvProp* **where** *appInvProp a I P* ≡
    $\forall x\ x'\ d\ p\ p'.\ I\ x\ \wedge\ appCompute\ a\ x\ p\ d\ x'\ p' \longrightarrow I\ x' \wedge P\ (x',\ p')$

Supporting theories, e.g., on queues and states (i.e., maps to values or queues) are developed that are used to prove properties of the semantics. For instance, the proof that the order of the thread initialisations is irrelevant uses a theory on "merging" states that provides commutative merging operators on states. In this paper we focus on the presentation of the semantics and do not have space to discuss supporting theory.

6 Related Work

This work is part of a broader effort by members of the AADL community to increase the scope and precision of the modeling language's semantic definition. Most closely related is the work by Hugues et al. [22] that provides a mechanization of AADL model structures and standardized property (attribute) sets in the Coq proof assistant. A key goal of this ongoing work is to organize and document modeling language features to support the AADL standard committee and stakeholders, and the lay a foundation for eventually incorporating the formalism in the standard itself. The paper [22] focuses on syntactic and structural elements, but an accompanying open source repository also includes a number of interesting extensions and auxiliary material such as connections to Coq-specified real-time scheduling algorithms and initial work on representing the semantics of AADL threads in a Coq formalization of the DEVS discrete event simulator input language syntax. Our work is complementary in that it concentrates less on the structural aspects of AADL, but more on formalizing the execution state and run-time services (while clarifying aspects of these presented in the standard) to enable traceability and eventual proofs of correctness of AADL-driven code generation and run-time libraries. Other differences include our formalization of application logic and our initial framework for property-based reasoning (which are not present in [22]). This provides the foundation for future proofs of soundness of AADL contract languages such as GUMBO [21] and for mapping contracts down into code (e.g., as needed to fully justify property-based based testing against AADL model-based contracts as presented in our recent work [18]). Finally, we have placed more of an emphasis on designing aspects of the formalism as abstractions that can be formally refined towards specific platform and middleware semantics. There is no barrier to a deeper merging of the lines of work in either Coq or Isabelle.

While the above work has focused on mechanizations of AADL in theorem provers, there are many other contributions to the formal specification, analysis, and verification of AADL models and annexes. These works, whether implicitly or explicitly, identify a target model checker or simulation framework with its own semantics, and then encode aspects of AADL into the semantics of the target framework. Some contributions focus on static semantics of models [4, 31] while others consider run-time behavior and use model translation to extract executable specifications from AADL models, e.g., [6, 7, 10, 16, 33]. Rolland et al.

[27] formalized aspects of AADL's coordination and timing behavior through the translation of AADL models into the Temporal Logic of Actions (TLA+) [25]. Many related works formalize a subset of AADL (e.g., for synchronous systems only) or focus on analyzing an aspect of the system, such as schedulability [30], behavioral [8,32], or dependability analyses [13].

7   Conclusion

The mechanized semantics that we have presented for AADL execution is substantial, and it provides a foundation for designing and verifying infrastructure for AADL-aligned model-driven development. Its immediate impact is realized in the HAMR framework, which is being used in a number of industrial research projects. It complements other recent work [22] on mechanized AADL semantics by providing specifications and proofs for run-time state, including formalizations of key aspects of AADL run-time services, as well as property frameworks for component-level and end-to-end system reasoning. A valuable contribution is the holistic integration of these semantic aspects (component and system state, contracts, run-time services) to better support clean design and integration of accompanying tooling including contract-based verification of components [21], property-based testing of components [18], and multi-platform code generation [17]. Our approach emphasizes setting up abstract definitions, e.g., for communication, contracts, scheduling, that can subsequently be specialized via formal refinement to definitions for particular platforms and deployment scenarios.

AADL is a large modeling language, and we treat only a subset – chosing to omit at present more complicated features such as AADL modes and error recovery. In particular, we hope to include in the near future notions related to timing to be able to provide an interpretation for AADL's timing-related properties and to justify contract language features for time. We have found it challenging enough to obtain a formalization for our selected features and align them with our experience in code generation for multiple platforms and our implementation of contract-based verification and testing. We believe that our scope is reasonable because it aligns with features that have been used to support US defense-related research projects at Collins Aerospace that use AADL and HAMR to implement, e.g., prototypes of subsystems for the CH-47 Chinook military helicopter platform [11]. One aspect of our current formalism includes showing how the static cyclic scheduling used on those projects could be incorporated with other dimensions of the semantics.

8   Future Work

This work lays for the foundation for several important next steps.

Mechanized Proofs of Soundness for GUMBO Contract Framework: In previous work, Hatcliff et al. developed the AADL contract language that enables compositional reason about AADL systems. GUMBO is supported by both SMT-based verification [21] and property-based testing [18] tools that can be used to demonstrate that component application code written in the Slang subset of Scala [26] conforms to GUMBO contracts. GUMBO is inspired by ideas from the AGREE [12] and BLESS [24] but puts a greater emphasis on aligning with the AADL run-time semantics and threading structure and with enabling application code to be verified against model-level component contracts. We are building the contract representations in AADL-HSM to formalize the semantics of GUMBO component contracts and to enhance the current notions of compositionality in GUMBO to support system-level properties and proofs. This includes both proving the soundness of the verification condition generation for SMT-based reasoning [21] as well as the generation of executable contracts used in the accompanying property-based testing framework [18]. We are extending the HAMR generation of Isabelle AADL-HSM artifacts to include translation of GUMBO contracts. Although this would allow system properties of AADL models to be proved directly in Isabelle, we believe that greater end-user usability will be achieved if we use experience with the Isabelle-based definitions to design and prove the soundness of a highly automated deduction framework for application logic composition that would be incorporated as an extension to the Logika verification framework [26] (allowing developers to work directly in industrial IDEs instead of in Isabelle). AADL models and contracts from the Open PCA Pump medical device project [20] are being used as a driver for this work. It also seems possible to use AADL-HSM to address soundness of AADL-level contracts for secure information flow [2,3].

Guidance for New Platform Backends: HAMR backends for seL4 and Linux are being refined and extended in ongoing industry projects. We are working on AADL-HSM extensions that specify the semantics of the AADL run-time at lower levels of abstraction corresponding to the different HAMR-based implementations on specific platforms. We intend that the AADL-HSM design will enable us to show these lower-level mechanization to be formal refinements of the general specifications. We are particularly interested over the long term in developing refinements to the Isabelle-based specifications of seL4. Our previous and ongoing collaborations with the seL4 team at ProofCraft and University of New South Wales on the DARPA CASE project are enabling preliminary planning for this effort [5].

Traceability and Correctness of HAMR Code Generation: Many aspects of HAMR code generation are already aligned with our semantics, e.g., one can inspect the data types for HAMR's port state and thread state and the implementation of AADL RTS and observe the correspondence with AADL-HSM specifications. We are continuing to refactor both HAMR code generation to achieve stronger traceability, as well as the ability to translate arbitrary HAMR run-time states (as captured via logging) into Isabelle. Our ultimate and rather ambitious aim is to prove the soundness of HAMR code generation with respect to AADL-HSM. This would require utilizing, e.g., the C mechanized semantics infrastructured used in the seL4 proof base. HAMR code generation is factored through Slang [26] (a safety-critical subset of Scala support by the Logika verifier). Some aspects of the HAMR AADL run-time, e.g., the thread dispatch logic are written in a purely functional subset of Slang, while other sections (e.g., run-time service implementations) are amenable to Logika specification and verification. As a nearer-term goal, we are investigating translating purely functional Slang into Isabelle functions and making a stronger connection between Logika specifications and verification conditions and our Isabelle definitions.

References

1. Architecture analysis and design language (AADL), SAE AS5506 Rev. C (2017)
2. Amtoft, T., et al.: A certificate infrastructure for machine-checked proofs of conditional information flow. In: Degano, P., Guttman, J.D. (eds.) POST 2012. LNCS, vol. 7215, pp. 369–389. Springer, Heidelberg (2012). https://doi.org/10.1007/978-3-642-28641-4_20
3. Amtoft, T., Hatcliff, J., Rodríguez, E., Robby, Hoag, J., Greve, D.: Specification and checking of software contracts for conditional information flow. In: Cuellar, J., Maibaum, T., Sere, K. (eds.) FM 2008. Formal Methods, vol. 5014, pp. 229–245. Springer, Heidelberg (2008). https://doi.org/10.1007/978-3-540-68237-0_17
4. Backes, J., Cofer, D., Miller, S., Whalen, M.W.: Requirements analysis of a quad-redundant flight control system. In: Havelund, K., Holzmann, G., Joshi, R. (eds.) NFM 2015. LNCS, vol. 9058, pp. 82–96. Springer, Cham (2015). https://doi.org/10.1007/978-3-319-17524-9_7
5. Belt, J., et al.: Model-driven development for the seL4 microkernel using the HAMR framework. J. Syst. Architect. **134**, 102789 (2022)
6. Berthomieu, B., Bodeveix, J.-P., Chaudet, C., Dal Zilio, S., Filali, M., Vernadat, F.: Formal verification of AADL specifications in the topcased environment. In: Kordon, F., Kermarrec, Y. (eds.) Ada-Europe 2009. LNCS, vol. 5570, pp. 207–221. Springer, Heidelberg (2009). https://doi.org/10.1007/978-3-642-01924-1_15
7. Berthomieu, B., et al.: Formal verification of AADL models with fiacre and tina. In: ERTSS 2010-Embedded Real-Time Software and Systems, pp. 1–9 (2010)
8. Besnard, L., et al.: Formal semantics of behavior specifications in the architecture analysis and design language standard. In: Nakajima, S., Talpin, J.-P., Toyoshima, M., Yu, H. (eds.) Cyber-Physical System Design from an Architecture Analysis Viewpoint, pp. 53–79. Springer, Singapore (2017). https://doi.org/10.1007/978-981-10-4436-6_3
9. Burns, A., Wellings, A.: Analysable Real-Time Systems: Programmed in Ada. CreateSpace (2016)
10. Chkouri, M.Y., Robert, A., Bozga, M., Sifakis, J.: Translating AADL into BIP - application to the verification of real-time systems. In: Chaudron, M.R.V. (ed.) MODELS 2008. LNCS, vol. 5421, pp. 5–19. Springer, Heidelberg (2009). https://doi.org/10.1007/978-3-642-01648-6_2
11. Cofer, D.D., et al.: Cyberassured systems engineering at scale. IEEE Secur. Priv. **20**(3), 52–64 (2022)
12. Cofer, D., Gacek, A., Miller, S., Whalen, M.W., LaValley, B., Sha, L.: Compositional verification of architectural models. In: Goodloe, A.E., Person, S. (eds.) NFM 2012. LNCS, vol. 7226, pp. 126–140. Springer, Heidelberg (2012). https://doi.org/10.1007/978-3-642-28891-3_13
13. Feiler, P., Rugina, A.: Dependability modeling with the architecture analysis and design language (AADL). Carnegie-Mellon University of Pittsburgh PA Software Engineering INST, Technical report (2007)

14. Feiler, P.H.: Efficient embedded runtime systems through port communication optimization. In: 13th IEEE International Conference on Engineering of Complex Computer Systems, pp. 294–300. IEEE (2008)
15. Feiler, P.H., Gluch, D.P.: Model-Based Engineering with AADL: An Introduction to the SAE Architecture Analysis & Design Language. Addison-Wesley, Boston (2013)
16. Hadad, A.S.A., Ma, C., Ahmed, A.A.O.: Formal verification of AADL models by event-B. IEEE Access **8**, 72814–72834 (2020)
17. Hatcliff, J., Belt, J., Robby, Carpenter, T.: HAMR: an AADL multi-platform code generation toolset. In: Margaria, T., Steffen, B. (eds.) ISoLA 2021. LNCS, vol. 13036, pp. 274–295. Springer, Cham (2021). https://doi.org/10.1007/978-3-030-89159-6_18
18. Hatcliff, J., Belt, J., Robby, Legg, J., Stewart, D., Carpenter, T.: Automated property-based testing from AADL component contracts. In: Cimatti, A., Titolo, L. (eds.) FMICS 2023. LNCS, vol. 14290, pp. 131–150. Springer, Cham (2023). https://doi.org/10.1007/978-3-031-43681-9_8
19. Hatcliff, J., Hugues, J., Stewart, D., Wrage, L.: Formalization of the AADL runtime services. In: Margaria, T., Steffen, B. (eds.) ISoLA 2022. LNCS, vol. 13702, pp. 105–134. Springer, Cham (2022). https://doi.org/10.1007/978-3-031-19756-7_7
20. Hatcliff, J., Larson, B.R., Carpenter, T., Jones, P.L., Zhang, Y., Jorgens, J.: The open PCA pump project: an exemplar open source medical device as a community resource. SIGBED Rev. **16**(2), 8–13 (2019)
21. Hatcliff, J., Stewart, D., Belt, J., Robby, Schwerdfeger, A.: An AADL contract language supporting integrated model- and code-level verification. In: Proceedings of the 2022 ACM Workshop on High Integrity Language Technology. HILT 2022 (2022)
22. Hugues, J., Wrage, L., Hatcliff, J., Stewart, D.: Mechanization of a large DSML: an experiment with AADL and coq. In: 20th ACM-IEEE International Conference on Formal Methods and Models for System Design, MEMOCODE 2022, pp. 1–9. IEEE (2022)
23. Klein, G., et al.: Comprehensive formal verification of an OS microkernel. ACM Trans. Comput. Syst. **32**(1), 1–70 (2014)
24. Larson, B.R., Chalin, P., Hatcliff, J.: BLESS: formal specification and verification of behaviors for embedded systems with software. In: Brat, G., Rungta, N., Venet, A. (eds.) NFM 2013. LNCS, vol. 7871, pp. 276–290. Springer, Heidelberg (2013). https://doi.org/10.1007/978-3-642-38088-4_19
25. Merz, S.: The specification language TLA$^+$. In: Bjørner, D., Henson, M.C. (eds.) Logics of Specification Languages. MTCSAES, pp. 401–451. Springer, Heidelberg (2008). https://doi.org/10.1007/978-3-540-74107-7_8
26. Robby, Hatcliff, J.: Slang: the sireum programming language. In: Margaria, T., Steffen, B. (eds.) ISoLA 2021. LNCS, vol. 13036, pp. 253–273. Springer, Cham (2021). https://doi.org/10.1007/978-3-030-89159-6_17
27. Rolland, J.F., Bodeveix, J.P., Chemouil, D., Filali, M., Thomas, D.: Towards a formal semantics for AADL execution model. In: Embedded Real Time Software and Systems (ERTS2008) (2008)
28. SAnToS Laboratory: HAMR project website (2022). https://hamr.sireum.org
29. SAnToS Laboratory: AADL HAMR semantics mechanization - git repository (2023). https://github.com/santoslab/AADL-HSM
30. Sokolsky, O., Lee, I., Clarke, D.: Schedulability analysis of AADL models. In: Proceedings 20th IEEE International Parallel & Distributed Processing Symposium, p. 8. IEEE (2006)

31. Stewart, D., Liu, J.J., Whalen, M., Cofer, D., Peterson, M.: Safety annex for architecture analysis design and analysis language. In: ERTS 2020: 10th European Conference Embedded Real Time Systems, p. 10 (2020)
32. Tan, Y., Zhao, Y., Ma, D., Zhang, X.: A comprehensive formalization of AADL with behavior annex. Sci. Program. **2022**, 2079880 (2022)
33. Yang, Z., Hu, K., Ma, D., Bodeveix, J.P., Pi, L., Talpin, J.P.: From AADL to timed abstract state machines: a verified model transformation. J. Syst. Softw. **93**, 42–68 (2014)

A Formal Web Services Architecture Model for Changing PUSH/PULL Data Transfer

Naoya Nitta(✉), Shinji Kageyama, and Kouta Fujii

Konan University, 8-9-1 Okamoto, Kobe, Japan
n-nitta@konan-u.ac.jp, m2124002@a.konan-u.ac.jp, m2224001@s.konan-u.ac.jp

Abstract. Deciding how data should be transferred among Web services is an important part of their architecture design. Basically, each piece of data is transferred in either PUSH or PULL style. The architect's selection of data transfer methods generally has a great impact on both the overall structure and performance of Web services. However, little work has been done on helping architects to select suitable data transfer methods. In this paper, we present a formal model to abstract some parts of Web services architecture that are not affected by the selection of data transfer methods, and based on the model, we propose an architecture level refactoring for changing data transfer methods. Also, we present an algorithm to generate prototypes of Web services from the architecture model and selected data transfer methods, and proved the correctness of the algorithm. Furthermore, we developed a tool that provides a graph-based UI for the refactoring and can generate executable prototypes of Web services. To evaluate our method, we conducted case studies for several Web applications and confirmed that the generated prototypes can be used to estimate the performance.

Keywords: Web services architecture model · PUSH/PULL data transfer · Architecture model refactoring

1 Introduction

Many kinds of data are created, changed and transferred among Web services. Between two Web services, each piece of data is basically transferred in either PUSH or PULL style. In PUSH-style, the source side service sends data to the destination side one as a parameter of an API call, but in PULL-style, the destination side one gets data from the source side one as a response of an API call. Generally, selection of data transfer methods has a great impact on both the overall structure and performance of Web services. However, little work has been done on helping architects to select suitable data transfer methods.

In this paper, we present a formal architecture model to support selection of data transfer methods among Web services. Before explaining the model, first, we think about the meaning of selection of data transfer methods. For example, consider the case of transferring data from service A to service B. If PUSH-style has been selected for the transfer, then B's API is called by A and the control flows in the same direction of the data-flow, but in PULL-style, A's API is called by B and the control flows in the reverse direction. Therefore, selection of a data transfer method corresponds to selection of the direction of the control-flow.

If some 'control-independent' parts of Web services architecture can be specified before selecting data transfer methods, then their selection and reselection tasks will be simplified. However, many formal architecture models [1, 6, 8] cannot be used for the purpose because they are basically process-centric and the process based descriptions are control-dependent. Here, by control-dependent, we mean that the model contains some information about control-flow and its descriptions are affected by a change of the structure of control-flow. Therefore in this paper, we introduce a formal architecture model in which control-dependent information is abstracted away. To make the model control-independent, each component of the model is assumed to have no temporal state and no internal transition. Therefore in the model, every state transition of every component is always observable and the states of all components are assumed to be changed synchronously. On the basis of the architecture model, selection of data transfer methods is regarded as addition of the information about the directions of control-flow to the model, and at least at the architecture level, the data transfer methods can be refactored simply by changing the added information. In this paper, we present an algorithm to generate a prototype of Web services from our architecture model and the added information, and prove the correctness of the algorithm. We also designed a simple architecture description language (ADL for short) to specify the architecture model, and developed a tool that can read a model file written in the ADL, provides a graph-based UI to select data transfer methods and can generate executable JAX-RS prototypes so that they are used to estimate the impact of selection of data transfer methods on the performance. To evaluate our method, we conducted case studies for several simple Web applications and confirmed that the generated prototypes can be used to estimate the performance.

2   Motivating Example

First, we show an example of Web services in which data transfer methods can be changed to PUSH or PULL-style. The subject services are simple weather observation services (in the following, WOS for short). For the simplicity, we assume that the services deal with only a single weather station. The functional requirements are as follows. Whenever the station observes the weather, the temperature in Fahrenheit is measured and that in Celsius is instantly calculated from the measured temperature. At the same time, the value of the highest temperature (in Fahrenheit) is updated using the measured temperature. In addition, the highest temperature is reset every time a date changes.

Here, assume that the services are planned to be constructed as RESTful [3] Web services and consist of three resources; the temperature in Fahrenheit (denoted by `temp_f`), the temperature in Celsius (denoted by `temp_c`) and the

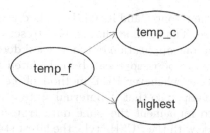

Fig. 1. Dataflow Graph of WOS System

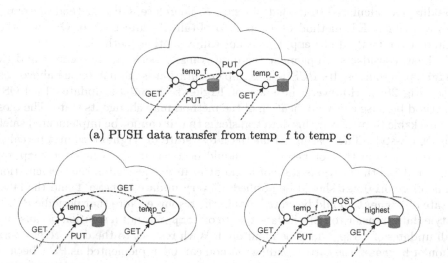

(a) PUSH data transfer from temp_f to temp_c

(b) PULL data transfer from temp_f to (c) PUSH data transfer from temp_f to
temp_c   highest

Fig. 2. Possible Implementations of Data Transfer in WOS System

highest temperature (denoted by **highest**). We show the data-flow among the
resources in Fig. 1. In a RESTful style of architecture, each resource corresponds
to a piece of information on the Web that can be identified by a URI. Each
resource can provide HTTP methods such as GET, PUT, POST and DELETE.
The GET method is assumed to be *safe*, which means that the state of the
resource is not changed by applying the method, and the GET, PUT and
DELETE methods are assumed to be *idempotent*, which means that its multiple applications to the resource cause the same effects on the resource as its
single application. Typically, the GET method is used to obtain the state of a
resource, and the PUT and POST methods are used to update the state of a
resource.

First, consider to implement the data transfer between the **temp_f** and
temp_c resources in PUSH-style (see Fig. 2(a)). For each resource, the latest

state is stored within the resource and its GET method is implemented to return the state as its response. The method that is used to set a new state to `temp_f` is PUT since more than one application of the method does not change its state. Here, recall that the state of `temp_c` can be calculated from that of `temp_f`. Therefore in PUSH-style, whenever PUT method of `temp_f` is called, it also calls PUT method of `temp_c` so that its internal state is kept up-to-date.

Second, consider to implement the same data transfer in PULL-style (see Fig. 2(b)). As is the case with the PUSH-style, the latest state of `temp_f` is stored within the resource and it is returned by calling its GET method. However in PULL-style, the latest state of `temp_c` is not stored anywhere because it can be directly calculated from that of `temp_f`. Therefore, GET method of `temp_c` always calls GET method of `temp_f` to obtain its latest state. On the other hand, PUT method of `temp_f` does not call any other method.

Last, consider to implement the data transfer between the `temp_f` and the `highest` resources. Its PUSH-style implementation is similar to the above case (see Fig. 2(c)). However, the state of `highest` should be updated by POST method because every application of the method may change its state. The most remarkable thing about this data transfer is that it cannot be implemented safely in PULL-style. The reason is that the latest state of `highest` cannot be calculated only from that of `temp_f`. It should be recalculated whenever `temp_f` is updated and the previous state of `highest` is always needed in the recalculation. Therefore, `highest` should be notified of every update of `temp_f` and the latest state of `highest` should be stored within the resource. Note that periodic PULL-style data transfer to poll the latest state of `temp_f` is not a solution because not all updates of `temp_f` may be monitored. With respect to the daily reset operation of `highest`, the corresponding method can be implemented as PUT because the operation satisfies idempotency. By comparing the above implementations, we can make the following observations.

- **Some data transfer method cannot be changed to PULL-style.** In the example of WOS, the data transfer from `temp_f` to `highest` cannot be implemented safely in PULL-style, but that from `temp_f` to `temp_c` can be. Just looking at the data-flow in Fig. 1, we cannot distinguish these properties of data transfer. To determine the property, we need to know the concrete relation between the source and destination resources of the data transfer. Therefore, in the next section, we present a formal model that can represent the relation among state changes of the relevant resources.
- **Some resource's state should be stored within the resource and the other ones do not need to be stored.** At the implementation level, the state of a resource may not be stored within the resource even if it is always observable from the outside of the system and its value can always be obtained. For example, the state of `temp_c` is not stored anywhere in the PULL-style implementation although it is stored within the resource in the PUSH-style implementation. We can say that the state of a resource should be stored within the resource if it is updated by PUSH-style transfer.

– **Data transfer methods will affect the non-functional properties of the services.** Selection of the data transfer methods is tightly associated with the non-functional properties of the services. For example in a PUSH-style implementation, since the states of the destination resources should be stored in the memory or storage, it will increase the consumption. The communication load to update the resource's state is higher in PUSH-style than in PULL-style, but that to observe the resource's state is the opposite.

3   Overview of Architecture Model

In this paper, we present a formal architecture model in which control-dependent information is abstracted away. In this section, we explain its overview. Its formal definition is given in the next section. As stated in Sect. 1, process based descriptions are generally control-dependent. For example, consider the case that process P sends message b in response to receiving message a. In process-based models like CCS [7], P is modeled as having a halfway state where it has received a but still has not sent b. From the state, by performing an internal action, which is unobservable from the outside, P sends b and becomes the next state. We think such a halfway state and internal action make the computation model control-dependent. Therefore in our model, each component is assumed to have no halfway state and no internal transition. In addition, every state transition of every component is observable and the states of all components are changed synchronously. In our model, a *resource* is a software component c that satisfies:

1. c is identifiable from the outside of the system,
2. c seems to have its own state and the state is always observable from the outside of the system,
3. the state of c changes only when c receives or sends some message.

Note that these are just architecture level assumptions. By adding control-dependent information to the model, each component can be implemented to have halfway states and internal transitions.

We illustrate our architecture model based on the WOS explained in the previous section. In our architecture model, a resource is modeled as a state transition system whose state is changed by receiving or sending a message. In several existing architecture models (e.g. [5]), each component is modeled as a labeled transition system (LTS). However, since the state space of LTS is limited to finite, it does not have sufficient expressive power to represent many kinds of data structure such as natural numbers and lists. In addition, LTS is not suitable for generating a readable program since the state changes will be translated as a large number of conditional branches. Therefore in our model, we do not limit the state space of transition systems to finite. For example, we model the state transition of `temp_f` by calling its PUT method with parameter x as:

$$\text{temp\_f}(f, x) = x.$$

Each state transition function takes a state of a resource and a message as its first and second arguments, and returns its next state. Similarly, the state transition functions to update `temp_c` and `highest` can be modeled as follows.

$$\text{temp\_c}(c, x) = (x - 32)/1.8.$$
$$\text{highest}(h, x) = \text{if}(x > h, x, h).$$

In these definitions, we directly use the name of each resource as the name of its state transition function, and assume that these resources receive the same message x synchronously.

The above formulas may seem too concrete to be used in an architecture model. However, as discussed in the previous section, without knowing the concrete relationship between the source and destination resources, it is impossible to know whether the data transfer method can be changed to PULL-style or not. For example, consider why the data transfer from `temp_f` to `temp_c` can be implemented in PULL-style. In the above formulas, x corresponds to a new state of `temp_f`. As we can see from the formulas, the next state of `temp_c` (i.e., $(x - 32)/1.8$) can be calculated only from x, but to calculate the next state of `highest` (i.e., if$(x > h, x, h)$), in addition to x, its current state h is also needed. This difference determines the implementability of data transfer in PULL-style. More generally, with respect to data transfer from resource s to resource d, if there exists a homomorphism that maps the state transition of s to that of d, then the data transfer can be implemented in PULL-style (see Sect. 5.3 for the detail). For example with respect to the data transfer from `temp_f` to `temp_c`, a function $h(x) = (x - 32)/1.8$ is a homomorphism since it satisfies:

$$h(\text{temp\_f}(f, x)) = h(x) = (x - 32)/1.8 = \text{temp\_c}(h(f), x).$$

By using h, the data transfer can be implemented in PULL-style. On the other hand, since there is no homomorphism from the state transition of `temp_f` to that of `highest`, it cannot be implemented in PULL-style. In this way, whether each data transfer can be implemented in PULL-style or not can be determined by comparing the state transition functions of the source and destination resources.

To make the source side and the destination side state transition functions comparable, in our architecture model, the formulas that define these functions are grouped by a *channel*. A channel is defined for each data transfer, and in each channel, we assume that the states of the source and destination resources are changed synchronously by the same message on the channel. Each channel is assumed to have three ports; the input, output and reference ports, and the source and destination resources are assumed to connect to its input and output ports, respectively. The state of a resource connected to the reference port does not change by the message on the channel. Semantically, these assumptions mean that a message is transferred from the input side and reference side resources to an output side one through the channel instantaneously, and the state change on the input side resource, the data transfer and the state change on the output side resource are all synchronized. Such assumptions make the architecture model capable of representing data-flow but independent of control-flow. Here, assume that data from temp_f to temp_c is transferred through a channel named c_1. Then, in our architecture model, the data transfer can be represented as:

$$\text{temp\_f}^{c_1,\text{I}}(f, y) = y,$$
$$\text{temp\_c}^{c_1,\text{O}}(c, y) = (y - 32)/1.8.$$

In these formulas, the superscript of each function consists of the channel and its port I, O or R that the resource connects to. In this data transfer, y is the transferred message from temp_f to temp_c on channel c_1 and the message is calculated from the next state of temp_f, which is the input side resource of c_1. In general, multiple resources can connect to the input and reference ports of a channel, and a message on the channel is calculated from the current and next states of the input side resources and the current states of the reference side resources. For example, consider the following channel c.

$$\text{pos\_x}^{c,\text{I}}(x, \langle dx, dy \rangle) = x + dx = x',$$
$$\text{pos\_y}^{c,\text{I}}(y, \langle dx, dy \rangle) = y + dy = y',$$
$$\text{distance}^{c,\text{O}}(d, \langle dx, dy \rangle) = d + \sqrt{dx^2 + dy^2} = d'.$$

The input side resources pos_x and pos_y represent the 2D position of something and the output side resource distance represents its travel distance. In this case, the message $\langle dx, dy \rangle$ is calculated from x, x', y and y' as $\langle x' - x, y' - y \rangle$. However, such a calculation may seem counterintuitive because the constraints that the calculation requires are a kind of inverse mapping of the input side state transition functions but the given constraints are in the form of state transition functions themselves. The reasons why we define such constraints for each channel separately and in the form of state transition functions are as follows.

- Since the massages sent on each channel generally differ depending on the channel, the input side constraints used to calculate the messages also differ depending on the channel, and should be defined separately.
- As stated before, to determine the changeability of the data transfer method to PULL-style, both the input side and output side constraints should be given in the state transition function form.
- When multiple resources connect to input and/or reference ports, as an inverse mapping, the constraints must be given in a multi valued inverse mapping for each resource (e.g., $x, x' \mapsto \langle x' - x, dy \rangle$ for pos_x), or a single valued inverse mapping of all input/reference side resources (e.g., $x, x', y, y' \mapsto \langle x' - x, y' - y \rangle$). However, both definitions also seem counterintuitive.
- If both the input side and output side constraints have the same form, then flipping their sides can flexibly be done.

The interactions between the system and its environment except for the observation of each resource's state are done through special channels for input events.

Table 1. Connection between Channels and Resources in WOS

Channel	Input side resource	Output side resource
c_{evt_1}		temp_f
c_{evt_2}		highest
c_1	temp_f	temp_c
c_2	temp_f	highest

$$\text{temp\_f}^{c_{evt_1},\text{O}}(f,x) = x,$$
$$\text{highest}^{c_{evt_2},\text{O}}(h,v) = v,$$
$$\text{temp\_f}^{c_1,\text{I}}(f,y) = y,$$
$$\text{temp\_c}^{c_1,\text{O}}(c,y) = (y-32)/1.8,$$
$$\text{temp\_f}^{c_2,\text{I}}(f,z) = z,$$
$$\text{highest}^{c_2,\text{O}}(h,z) = \text{if}(z > h, z, h).$$

Fig. 3. Data Transfer Architecture Model of WOS

The observation of the state is assumed to be done directly from the environment without going through any channel.

In the following, we explain how the whole WOS can be modeled as an architecture model. In this example, we use channels c_{evt_1}, c_{evt_2}, c_1 and c_2, and let c_{evt_1} and c_{evt_2} be event channels. We show the connection between the resources and the channels in Table 1. The state transition functions of the resources are defined in Fig. 3. In our model, the state transition function of the same resource seems to be multiply defined in different channels, but they are just different images of the same function. Therefore, the first arguments of them and their function values are always identical, respectively. For example in Fig. 3, always $x = y = z$ holds. The whole system reacts in response to a message from the environment through an event channel. For example, if message 68 is input to the WOS through channel c_{evt_1}, then $\text{temp\_f}^{c_{evt_1},\text{O}}(f, 68)$ is evaluated and the next state of temp_f is shown to be 68. Since $\text{temp\_f}^{c_{evt_1},\text{O}}$, $\text{temp\_f}^{c_1,\text{I}}$ and $\text{temp\_f}^{c_2,\text{I}}$ are different images of the same function, we have $x = y = z = 68$. By these equations, both messages sent on c_1 and c_2 become 68. Finally, the states of temp_c and highest respectively become 20 and 68 if the previous state of highest is less than 68.

4   Data Transfer Architecture Model

4.1   Basic Definitions

We define a data transfer architecture model as $\mathcal{R} = \langle R, C, \rho, D, \tau, \mu, \Delta, s_0 \rangle$ where:

- R: a finite set of *resources*,
- C: a finite set of *channels*,
- ρ ($\rho : C \times \{\mathrm{I}, \mathrm{O}, \mathrm{R}\} \to 2^R$): a *connection map* that maps a channel and its port (input, output or reference) to the set of all connecting resources,
- D: a finite/infinite set of data values,
- τ ($\tau : R \to 2^D$): a map from a resource to the set of its all states,
- μ ($\mu : C \to 2^D$): a map from a channel to the set of all messages that can be transferred on the channel,
- $\Delta = \langle \delta_r^{c,d} \rangle_{c \in C, d \in \{\mathrm{I},\mathrm{O},\mathrm{R}\}, r \in \rho(c,d)}$: a finite set of state transition functions,
- s_0: the initial state ($s_0(r) \in \tau(r)$ for each $r \in R$).

The details of the model will be explained in the following.

4.2   Resource and Channel

A resource is a component that seems to have its own state and the state is always observable from the outside of the system. The set of all resources within the system is denoted by R. In this paper, we assume that the number of all resources is fixed and no resource is either created or deleted at runtime. However, each resource can have an infinite state space. A channel is used to synchronize the state changes in relevant resources. The set of all channels in the system is denoted by C. Each channel in C can have one input, one output and one reference ports, and to each port, an arbitrary number of resources can connect. For each channel $c \in C$, $\rho(c, \mathrm{I})$, $\rho(c, \mathrm{O})$ and $\rho(c, \mathrm{R})$ represent the sets of all resources that connect to the input, output and reference ports of c, respectively. These sets are assumed to be disjoint. For simplicity, we call $r \in \rho(c, \mathrm{I})$, $r' \in \rho(c, \mathrm{O})$ and $r'' \in \rho(c, \mathrm{R})$ an input side, output side and reference side resources of c, respectively. If the state of an input side resource of a channel c is changed, then a message m is sent on c and the state of an output side resource of c is changed by receiving m. C always contains non-empty set C_{evt}. Every event channel $c_{\mathrm{evt}} \in C_{\mathrm{evt}}$ has no input side resource, that is, $\rho(c_{\mathrm{evt}}, \mathrm{I}) = \emptyset$.

4.3   States and Messages

In our architecture model, a data value is either a state of a resource or a message on a channel. The set of all data values is denoted by D. For each resource $r \in R$, $\tau(r)$ represents the set of all possible states of r, and for each channel $c \in C$, $\mu(c)$ represents the set of all possible messages that can be transferred on c. For each $c \in C$, an identity element $e_c \in \mu(c)$ that represents no operation is defined.

4.4   State Transition Functions

Δ is a finite set of state transition functions. It contains exactly one function for each connection of a resource to a channel. If a resource r connects to d side of a channel c (that is, $r \in \rho(c, d)$), then we denote the corresponding state transition function by $\delta_r^{c,d} : \tau(r) \times \mu(c) \to \tau(r)$. Each function takes a previous state of

r and a message on c as the first and second arguments, and returns the next state of r. For each $c \in C$ and $r \in R$, $\delta_r^{c,O}$ is always total and single valued, $\delta_r^{c,I}$ may be partial and/or multi valued, and $\delta_r^{c,R}$ is always multi valued and may be partial. Especially for $\delta_r^{c,I}$ and $\delta_r^{c,R}$, a more explanation would be needed. If $\delta_r^{c,R}(x, y) = z$ holds, then the value of y is always uniquely determined from the value of x regardless of the value of z, and thus $\delta_r^{c,R}$ is multi valued and may be partial. On the other hand, if $\delta_r^{c,I}(x, y) = z$ holds, then there can be some constraint on y and z such that the value of y is uniquely determined from the value of z, and thus $\delta_r^{c,I}$ may be multi valued and may be partial. For example, consider $\delta_{history}^{c,I}(h, \max(h2)) = h2$ where history is a resource whose state is a list of something and $\max(x)$ is the maximum value of the elements in list x. In this case, $h2$ is not uniquely determined from h and $\max(h2)$, and $\delta_{history}^{c,I}$ is multi valued. For a channel $c \in C$, let $\{r_1, \ldots, r_l\} = \rho(c, I)$, and $\{r_1', \ldots, r_k'\} = \rho(c, R)$. Then, to guarantee that a message on c can always be calculated from the current and next states of input side resources and the current states of reference side resources, there must be a *message generation function* $\delta^{-1,c}$ that satisfies for any $s_{r_1} \in \tau(r_1), \ldots, s_{r_l} \in \tau(r_l), s_{r_1'} \in \tau(r_1'), \ldots, s_{r_k'} \in \tau(r_k')$,

$$m = \delta^{-1,c}(s_{r_1}, \delta_{r_1}^{c,I}(s_{r_1}, m), \ldots, s_{r_l}, \delta_{r_l}^{c,I}(s_{r_l}, m), s_{r_1'}, \ldots, s_{r_k'}). \tag{1}$$

4.5   Dataflow Graph and Validity of Architecture Model

For a given data transfer architecture model $\mathcal{R}$, a *dataflow graph* is a directed graph $G_\mathcal{R} = (N_R, N_C, E)$ such that $N_R = R$, $N_C = C$ and $E = \{\langle c, r \rangle \mid c \in C, r \in \rho(c, O)\} \cup \{\langle r, c \rangle \mid c \in C, r \in \rho(c, I)\}$. For each event channel $c \in C_{\mathrm{evt}}$, a *dynamic dataflow graph* is the maximum subgraph $G_{\mathcal{R},c}$ of $G_\mathcal{R}$ such that every node in $G_{\mathcal{R},c}$ is reachable from node $c \in N_C$. We say $\mathcal{R}$ is *valid* when all of the following conditions are satisfied.

- $G_\mathcal{R}$ has no strongly connected component.
- For any event channel $c_{\mathrm{evt}} \in C_{\mathrm{evt}}$, $G_{\mathcal{R},c_{\mathrm{evt}}}$ is a directed tree.
- For any channel $c \in C$ and $r \in \rho(c, R)$, if r is contained in $G_{\mathcal{R},c_{\mathrm{evt}}}$, then r is a descendant of c on $G_{\mathcal{R},c_{\mathrm{evt}}}$.

4.6   State Transition of the Whole System

A state of the whole system is a composition of states of all resources. We model a state s of the whole system as a map from the resources to their states. That is, for each resource $r \in R$, s satisfies $s(r) \in \tau(r)$. Also the initial state s_0 of the system satisfies the same condition.

The whole system reacts in response to a message from the environment through an event channel. More specifically, if the state of $\mathcal{R}$ changes from s to s' by receiving a message m on $c_{\mathrm{evt}} \in C_{\mathrm{evt}}$, then we write $s \xrightarrow[\mathcal{R}]{\langle m, c_{\mathrm{evt}} \rangle} s'$, and the state change is possible if and only if there exists a *message assignment* $\pi : C \to D$ such that:

- for every $c \in C$, $\pi(c) \in \mu(c)$,
- $\pi(c_{\text{evt}}) = m$,
- for each $c \in C$, if c is contained in $G_{\mathcal{R},c_{\text{evt}}}$, then for $d \in \{I, O, R\}$ and $r \in \rho(c, d)$,

$$\delta_r^{c,d}(s(r), \pi(c)) \begin{cases} = s'(r) & (\delta_r^{c,d} \text{ is single valued}) \\ \ni s'(r) & (\delta_r^{c,d} \text{ is multi valued}), \end{cases} \tag{2}$$

- for each $c \in C$, if c is not contained in $G_{\mathcal{R},c_{\text{evt}}}$, then $\pi(c) = e_c$ and $s(r) = s'(r)$ for each r such that:
  - $r \in \rho(c, I)$, or
  - $r \in \rho(c, O)$ and r is not contained in $G_{\mathcal{R},c_{\text{evt}}}$.

For example, with respect to the architecture model explained in Sect. 3, $\{\text{temp\_f} \mapsto 66.2, \text{temp\_c} \mapsto 19, \text{highest} \mapsto 67.1\} \xrightarrow[\mathcal{R}]{\langle 68, c_{\text{evt}_1} \rangle} \{\text{temp\_f} \mapsto 68, \text{temp\_c} \mapsto 20, \text{highest} \mapsto 68\}$ holds since c_{evt_2} is not contained in $G_{\mathcal{R},c_{\text{evt}_1}}$ and there exists a message assignment $\pi = \{c_{\text{evt}_1} \mapsto 68, c_{\text{evt}_2} \mapsto e_{c_{\text{evt}_2}}, c_1 \mapsto 68, c_2 \mapsto 68\}$. In the following, for an input sequence $\sigma = \langle m_1, c_1 \rangle \cdots \langle m_n, c_n \rangle$, we write a concatenation of transition relations $\xrightarrow[\mathcal{R}]{\langle m_1, c_1 \rangle} \cdots \xrightarrow[\mathcal{R}]{\langle m_n, c_n \rangle}$ as $\xRightarrow{\sigma}_{\mathcal{R}}$.

5 RESTful Web Services Generation from Architecture Model and Selected Data Transfer Methods

As stated in Sect. 3, our architecture model is designed to be control-independent and not to be affected by changes of data transfer methods. Therefore, based on the architecture model, selection of data transfer methods is regarded as addition of the information about control-flow to the model. By adding such information, a prototype of RESTful Web services can be generated. In this section, we show a method to generate executable Web services from a data transfer architecture model and data transfer methods selected by a user. The generated prototype may be used to estimate the impact of selection of data transfer methods on the services performance.

5.1 Common Structure of RESTful Web Services for PUSH and PULL Data Transfer

As RESTful Web services, we consider to generate an executable prototype of JAX-RS Web application. JAX-RS is a Java API for RESTful Web services. Generally, a JAX-RS application consists of multiple resource classes, which are Java classes with some annotations. Let $\mathcal{R} = \langle R, C, \rho, D, \tau, \mu, \Delta, s_0 \rangle$ be an arbitrary data transfer architecture model. Since $\mathcal{R}$ has no information about control-flow, selection of data transfer methods is required to generate a JAX-RS prototype. Let E^{PLL} be the set of all data transfer methods where PULL-style are selected, that is, $E^{\text{PLL}} \stackrel{\text{def}}{=} \{\langle r, c \rangle \mid r \in R, c \in C, r \in \rho(c, I), \text{ and the data transfer method between } r \text{ and } c \text{ is PULL-style}\}$. Basically, from $\mathcal{R}$, a JAX-RS prototype $\mathcal{P}_{\mathcal{R}}$ is generated by mapping R to resource classes. For each resource $r \in R$, let P_r be the resource class corresponding to r. We define $\mathrm{In}_C(r) \overset{\text{def}}{=} \{c \in C \mid r \in \rho(c, \mathrm{O})\}$ and $\mathrm{Out}_C(r) \overset{\text{def}}{=} \{c \in C \mid r \in \rho(c, \mathrm{I})\}$. Also for any subset $C' \subseteq C$ of channels, we define $\mathrm{In}_R(C') \overset{\text{def}}{=} \bigcup_{c \in C'} \rho(c, \mathrm{I})$ and $\mathrm{Out}_R(C') \overset{\text{def}}{=} \bigcup_{c \in C'} \rho(c, \mathrm{O})$. Then, for any resource $r \in R$, P_r always has one getter (GET), $|\mathrm{In}_C(r) \cap C_{\text{evt}}|$ input (PUT/POST), and at most $|\mathrm{In}_R(\mathrm{In}_C(r) \backslash C_{\text{evt}})|$ update (PUT/POST) methods.

5.2   Generation of PUSH-First RESTful Web Services

Let $\mathcal{R}$ be an arbitrary data transfer architecture model. For any data transfer method in $\mathcal{R}$, we can always select PUSH-style. The JAX-RS prototype that is generated by selecting PUSH-style for every data transfer method (i.e., $E^{\text{PLL}} = \emptyset$) is called *PUSH-first prototype* and written as $\mathcal{P}_{\mathcal{R}}^{\text{PSH}}$. In this subsection, we will explain how to generate $\mathcal{P}_{\mathcal{R}}^{\text{PSH}}$ from $\mathcal{R}$.

First, for each resource $r \in R$, P_r always has a *state field*, that is a field to store its own state. The getter method of P_r is defined as follows.

```
public TypeOf_r get() {
    return this.stateOf_r;   // the state field of r
}
```

Next, let $\{r_1, \ldots, r_l\} = \mathrm{In}_R(\mathrm{In}_C(r))$. Then, for each j $(1 \leq j \leq l)$, P_r has a *cache field* of r_j, that is a field to store the cache of the latest state of r_j.

Last, we define the update and input methods of P_r. Let $\{r'_1, \ldots, r'_k\} = \mathrm{Out}_R(\mathrm{Out}_C(r))$ and consider each channel $c \in \mathrm{In}_C(r)$. If $c \notin C_{\text{evt}}$, then $|\mathrm{In}_R(\{c\})|$ update methods are defined for c. More concretely, for each $r_j \in \mathrm{In}_R(\{c\})$ (here, j satisfies $1 \leq j \leq l$), an update method of P_r for r_j is defined as follows.

```
public void update_from_rj(TypeOf_rj rj) {
    this.stateOf_r = δc,O(this.stateOf_r,
        δ-1,c(this.cacheOf_r1, this.cacheOf_r1,...,this.cacheOf_rj,
rj,...));
    this.cacheOf_rj = rj;
    r'1.update_from_r(this.stateOf_r);
        ⋮
    r'k.update_from_r(this.stateOf_r);
}
```

If $c \in C_{\text{evt}}$, then the input method for c is defined as follows.

```
public void input_on_c(TypeOf_m m) {
    this.stateOf_r = δc,O(this.stateOf_r,m);
    r'1.update_from_r(this.stateOf_r);
        ⋮
    r'k.update_from_r(this.stateOf_r);
}
```

5.3 Generation of PULL-Containing Web Services

In this subsection, we consider to change each data transfer method in $\mathcal{P}_{\mathcal{R}}^{\mathrm{PSH}}$ to PULL-style. Let r be a resource. For convenience of explanation, we assume that $\mathrm{Out}_C(r) = \{c'\}$ for some $c' \in C$ and let $\{r'_1, \ldots, r'_k\} = \mathrm{Out}_R(\{c'\})$. If $\langle r, c' \rangle \in E^{\mathrm{PLL}}$, then the generated prototype should satisfy the following conditions.

First, each resource class $P_{r'_i}$ ($1 \le i \le k$) has no state field and no update method. The getter method of $P_{r'_i}$ is redefined as:

```
public TypeOf_r'_i get() {
    return f_{r'_i}(..., r.get(), ...);
}
```

where $f_{r'_i}$ is a homomorphism that calculates the latest state of r'_i and will be explained below. $P_{r'_i}$ has no cache field of r.

Next, each update method of P_r is redefined as follows.

```
public void update_from_r_j(TypeOf_r_j r_j) {
    this.stateOf_r = δ_r^{c,O}(this.stateOf_r,
        δ^{-1,c}(this.cacheOf_r_1, this.cacheOf_r_1,...,this.cacheOf_r_j,
r_j,...));
    this.cacheOf_r_j = r_j;
}
```

By changing more than one data transfer method in $\mathcal{P}_{\mathcal{R}}^{\mathrm{PSH}}$ to PULL-style as the above, we can generate a *PULL-containing prototype* $\mathcal{P}_{\mathcal{R}}$. However, as discussed in Sect. 2, not all data transfer methods can be safely changed to PULL-style. In the following, we call the prototype in which as many data transfer methods as possible are changed to PULL-style *PULL-first*. In the next section, we will prove that $\mathcal{P}_{\mathcal{R}}^{\mathrm{PSH}}$ and any PULL containing prototype $\mathcal{P}_{\mathcal{R}}$ are equivalent if $\mathcal{P}_{\mathcal{R}}$ and $\mathcal{R}$ satisfy the following conditions.

Condition 1. *For every resource $r \in R$ and channel $c \in \mathrm{Out}_C(r)$, if $\langle r, c \rangle \in E^{\mathrm{PLL}}$, then for every resource $r' \in \mathrm{Out}_R(\{c\})$ and channel $c' \in \mathrm{Out}_C(r')$,*

1. $\langle r', c \rangle \in E^{\mathrm{PLL}}$,
2. $\mathrm{In}_C(r') = \{c\}$, and
3. $\rho(c, \mathrm{R}) = \emptyset$ holds.

Condition 2. *For every channel $c \in C$, let $\{r_1, \ldots, r_l\} = \mathrm{In}_R(\{c\})$ and if $\langle r_j, c \rangle \in E^{\mathrm{PLL}}$ for some j ($1 \le j \le l$), then for each resource $r' \in \mathrm{Out}_R(\{c\})$,*

$$s_0(r') = f_{r'}(s_0(r_1), \ldots, s_0(r_l)), \tag{3}$$

and for any state s of $\mathcal{R}$ and any message $m \in \mu(c)$ on c,

$$\delta_{r'}^{c,O}(f_{r'}(s(r_1), \ldots, s(r_l)), m) = f_{r'}(\delta_{r_1}^{c,I}(s(r_1), m), \ldots, \delta_{r_l}^{c,I}(s(r_l), m)). \tag{4}$$

This means that there exists a homomorphism $f_{r'}$ from the state transition systems of the source resources of c to that of each destination resource of c.

6 Equivalence of RESTful Web Services and Data Transfer Architecture Model

In this section, we give a brief outline of the proofs of the equivalence between a data transfer architecture model and any JAX-RS prototype generated from the model (for the details of the proofs, see the online appendix[1]). In advance of the proofs, first, we define some notations. Let $\mathcal{R}$ be an arbitrary data transfer architecture model and $\mathcal{P_R}$ be a JAX-RS prototype generated from $\mathcal{R}$. For $\mathcal{P_R}$, a *communication* $m(r, x_1, \ldots, x_n)/v$ is a pair of a request $m(r, x_1, \ldots, x_n)$ and its response v where m is a method of a resource r and $x_1, \ldots, x_n$ are parameters. Also for $\mathcal{P_R}$, we consider a communication sequence $\langle \widehat{m, c_{\text{evt}}} \rangle$ that corresponds to a message $\langle m, c_{\text{evt}} \rangle$ for $\mathcal{R}$. Let $\{r_1, \ldots, r_k\} = \rho(c_{\text{evt}}, O)$. Then,

$$\langle \widehat{m, c_{\text{evt}}} \rangle \overset{\text{def}}{=} \text{input\_on\_}c_{\text{evt}}(r_1, m)/\text{void} \cdots \text{input\_on\_}c_{\text{evt}}(r_k, m)/\text{void}.$$

In addition, let $\hat{\sigma} = \langle \widehat{m_1, c_1} \rangle \cdots \langle \widehat{m_n, c_n} \rangle$ if $\sigma = \langle m_1, c_1 \rangle \cdots \langle m_n, c_n \rangle$ is an input sequence for $\mathcal{R}$. For any state s of $\mathcal{R}$, we write a state of $\mathcal{P_R}$ as $\mathcal{P_R}(s)$ if for each resource class P_r of a resource r, $s(r)$ is stored in the state field in P_r and $s(r')$ is stored in any cache field of r' in P_r. If the state of $\mathcal{P_R}$ can be changed from $\mathcal{P_R}(s)$ to $\mathcal{P_R}(s')$ by a communication $m(r, x_1, \ldots, x_n)/v$, then we write $\mathcal{P_R}(s) \overset{m(r, x_1, \ldots, x_n)/v}{\longrightarrow} \mathcal{P_R}(s')$. For a communication sequence $\omega = \alpha_1 \cdots \alpha_n$, we write a concatenation of transition relations $\overset{\alpha_1}{\rightarrow} \cdots \overset{\alpha_n}{\rightarrow}$ as $\overset{\omega}{\Rightarrow}$.

Here, we prove the equivalence between a data transfer architecture model and the generated PUSH-first prototype.

Theorem 1. *Let* $\mathcal{R} = \langle R, C, \rho, D, \tau, \mu, \Delta, s_0 \rangle$ *be an arbitrary valid data transfer architecture model. Then, for any input sequence* σ, $\mathcal{P}_{\mathcal{R}}^{\text{PSH}}(s_0) \overset{\hat{\sigma}}{\Rightarrow} \mathcal{P}_{\mathcal{R}}^{\text{PSH}}(s)$ *iff* $s_0 \underset{\mathcal{R}}{\overset{\sigma}{\Rightarrow}} s$.

Proof Sketch. The theorem is proved by the definition of $\mathcal{P}_{\mathcal{R}}^{\text{PSH}}$ and induction on the number $n = |R|$ of the resources. For the induction step, $R' = R \backslash \{r\}$ (where $\text{Out}_C(r) = \emptyset$) is selected, and that the value stored in the state field of the resource class of r always equals to $s(r)$ is shown.   □

Next, we prove the equivalence between the PUSH-first prototype and any PULL-containing prototype generated from the same model.

Theorem 2. *Let* $\mathcal{R} = \langle R, C, \rho, D, \tau, \mu, \Delta, s_0 \rangle$ *be an arbitrary valid data transfer architecture model, and* $\mathcal{P_R}$ *be any JAX-RS prototype generated from* $\mathcal{R}$ *and satisfying conditions 1 and 2. Then,* $\mathcal{P}_{\mathcal{R}}^{\text{PSH}}$ *and* $\mathcal{P_R}$ *satisfy*

$$\forall_\sigma. \forall_{r \in R}. \{s_r \mid \exists_s. \mathcal{P}_{\mathcal{R}}^{\text{PSH}}(s_0) \overset{\sigma}{\Rightarrow} \mathcal{P}_{\mathcal{R}}^{\text{PSH}}(s) \overset{\text{get}(r)/s_r}{\Longrightarrow} \mathcal{P}_{\mathcal{R}}^{\text{PSH}}(s)\}$$

$$= \{s_r' \mid \exists_{s'}. \mathcal{P_R}(s_0) \overset{\sigma}{\Rightarrow} \mathcal{P_R}(s') \overset{\text{get}(r)/s_r'}{\Longrightarrow} \mathcal{P_R}(s')\}.$$

---

[1] https://nitta-lab.github.io/appendix_FACS2023.

Proof Sketch. The theorem is proved by double induction on the length of σ and the number $n = |R|$ of the resources. For the induction on n, also $R' = R\backslash\{r\}$ (where $\text{Out}_C(r) = \emptyset$) is selected, and that the responses of the getter methods of r in $\mathcal{P}_\mathcal{R}^{\text{PSH}}$ and that in $\mathcal{P}_\mathcal{R}$ are always equal is shown.   □

Last, we prove the equivalence between a data transfer architecture model and any generated JAX-RS prototype.

Theorem 3. *Let $\mathcal{R} = \langle R, C, \rho, D, \tau, \mu, \Delta, s_0 \rangle$ be an arbitrary valid data transfer architecture model, and $\mathcal{P}_\mathcal{R}$ be any JAX-RS prototype generated from $\mathcal{R}$ and satisfying conditions 1 and 2. For an arbitrary input sequence σ, $s_0 \overset{\sigma}{\underset{\mathcal{R}}{\Longrightarrow}} s$ holds if and only if there exists a state s' of $\mathcal{P}_\mathcal{R}$ such that $\mathcal{P}_\mathcal{R}(s_0) \overset{\hat{\sigma}}{\Rightarrow} \mathcal{P}_\mathcal{R}(s') \overset{\text{get}(r)/s(r)}{\Longrightarrow} \mathcal{P}_\mathcal{R}(s')$ holds for any resource $r \in R$.*

Proof. The theorem follows from Theorems 1 and 2.   □

7 Architecture Level Refactoring for Changing Data Transfer Methods

As stated in the previous section, from any valid data transfer architecture model $\mathcal{R}$, the PUSH-first JAX-RS prototype $\mathcal{P}_\mathcal{R}^{\text{PSH}}$ can be directly generated. Also, any PULL-containing JAX-RS prototype $\mathcal{P}_\mathcal{R}$, in which more than one data transfer method is changed to PULL-style can be generated from $\mathcal{P}_\mathcal{R}^{\text{PSH}}$. In this section, as an architecture level refactoring, we consider regenerating the PUSH-first or a PULL-containing JAX-RS prototype directly from $\mathcal{R}$. In the refactoring process, a user is required to select one combination of data transfer methods from all safe ones that satisfy the conditions shown in Sect. 5.3. The whole process of the refactoring is summarized as follows.

1. Check the validity of $\mathcal{R}$.
2. For each data transfer, PULL-style is determined to be selectable if the conclusion parts of condition 1-(2), condition 1-(3) and condition 2 are satisfied, and every data transfer method is initialized to PUSH-style.
3. Ask the user if he/she wants to change each data transfer method to PULL-style if possible. Throughout the selection process, condition 1-(1) is always checked, and the selection that violates the condition is not asked.
4. Generate a JAX-RS prototype based on the algorithm shown in Sect. 5.
5. Back to step 3 if the user wants to reselect other data transfer methods.

Among the above steps, only the check of the conclusion part of condition 2 in step 2 is considered difficult because determining the existence of a homomorphism is generally hard. Therefore in this paper, we consider to check a sufficient condition, *right unaryness* of the input and output side state transition functions. A function f is called right unary if f satisfies

$$f(x, z) = f(y, z) \tag{5}$$

for any x and y. The sufficiency of the right unaryness can be shown as follows. Let c be a channel, and assume that $\{r_1, \ldots, r_l\} = \rho(c, \mathrm{I})$ and $r' \in \rho(c, \mathrm{O})$. If for every i ($1 \leq i \leq l$), $\delta_{r_i}^{c,\mathrm{I}}$ is right unary, then there exists a function δ^{-1} such that:

$$m = \delta^{-1}(\delta_{r_1}^{c,\mathrm{I}}(s_1, m), \ldots, \delta_{r_l}^{c,\mathrm{I}}(s_l, m)),$$

and if $\delta_{r'}^{c,\mathrm{O}}$ is also right unary, then a function $f_{r'}$ that satisfies

$$f_{r'}(s_1', \ldots, s_l') = \delta_{r'}^{c,\mathrm{O}}(z, \delta^{-1}(s_1', \ldots, s_l'))$$

for any constant z becomes homomorphic and satisfies Eq. (4).

8   Tool Implementation

We defined a simple architecture description language (ADL for short) based on *many-sorted algebra* [4], and developed a graph-based refactoring tool[2] shown in Fig. 4. The tool first reads a model file written in the ADL and performs the steps 1 and 2 of the process shown in the previous section. Then, a data-flow graph is displayed on the tool and the user is allowed to select a safe combination of data transfer methods through pull-down menus. Last, the tool generates an executable JAX-RS prototype. In JAX-RS prototype generation, if the state transition function of a resource is right unary, then an API to update the resource state is implemented as PUT method otherwise POST method.

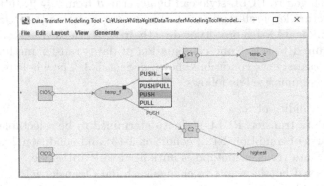

Fig. 4. Graph-based Tool To Refactor Data Transfer Methods

9   Case Studies and Discussion

9.1   Case Studies

As case studies, we designed the architectures of the following three applications and wrote the model files in the ADL.

---

[2] https://github.com/nitta-lab/DataTransferModelingTool.

- **WOS (Weather Observation Services)**: As a case study, we used the WOS explained in Sect. 3.
- **Inventory Management Services**: As another case study, we wrote a model file for inventory management services for a liquor store. The services take receiving reports and shipping requests as input and keep track of warehouse inventory and the waiting list.
- **Online Card Game**: The last one is an online game based on Algo. In Algo, each player is dealt a certain number of uniquely numbered cards. The cards are laid face-down and each player is required to guess the number of another player's card in turn. The player who is guessed all the cards loses the game.

Table 2. Total lines of model files and generated JAX-RS prototypes

Application	# of resources	Model File	PUSH-first	PULL-first
WOS	3	14	77	73
Inventory Management	7	43	361	346
Online Card Game	15	113	635	577

We generated the PULL-first/PUSH-first JAX-RS prototypes. The number of resources and the total lines of the model files and generated prototypes are summarized in Table 2. To confirm the effectiveness of the generated prototypes for performance estimation, we measured the computation time of each API call on the generated PULL-first/PUSH-first prototypes of the WOS. For the experiments, we deployed each JAX-RS prototype and a client program on the same PC[3], and measured the time to perform 10,000 iterations of API call five times and calculated the averaged value. We show the results in Fig. 5.

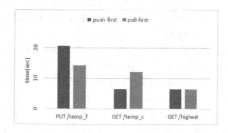

Fig. 5. Computation Time on JAX-RS Prototypes of WOS

---

[3] CPU: E5-1603 v4 2.80 GHz, RAM: 32.0 GB, JVM: jdk-12, Spring Boot: v2.4.1.

9.2   Discussion

As shown in Table 2, the generated prototypes were almost five times larger than the original descriptions. Furthermore, the implementations of the PULL-first and PUSH-first prototypes were significantly different. This suggests that addition of the control-dependent information can bridge a certain part of the gap between the abstraction levels of the architecture models and their implementations. In Fig. 5, the communication load to update a state is higher in PUSH-style than PULL-style, but that to obtain the state is higher in PULL-style than PUSH-style. These results conform to the tendency of the performance expected in Sect. 2. Therefore, we can expect that the generated prototypes can be used to estimate the impact of the selection on the resulting services performance.

10   Related Work

Little work has been done on helping architects to select suitable data transfer methods. In [12], Zhao presented a model of computation for push/pull communications. Compared to our model, the model has a less theoretical basis and does not cover components' behavior. However, some constraint on push and pull combinations (which is discussed in Sect. 5.3) was also discussed in the paper. In the context of parallel graph computation, Besta et al. [2] exhaustively analyzed the performance of push/pull variants of various graph algorithms.

Most of the formal architecture models (cf. [1,6,8]) are process-centric. The abstraction level of process-centric models is considered lower than ours because process-centric models are control-dependent. In fact, by adding the information about the control-flow to our model, CCS processes can be derived and the equivalence of derived processes can also be proved. In [5], labeled transition systems (LTLs) are used to represent and check behavior of RESTful Web applications. Since the state space of LTL is limited to finite, many temporal properties of the model can be automatically verified, but it does not have sufficient expressive power to represent many kinds of data structure such as natural numbers and lists. In contrast, our model can have infinite state space and can generate executable prototypes.

Dataflow programming [9–11] has been studied for about five decades. A dataflow program internally uses a directed graph that represents the set of all instructions and dataflow among the instructions. Both the representation and the semantics of dataflow programming are similar to ours, and in fact, our model can be implemented as a dataflow program rather straightforwardly. However, the abstraction level of our model is higher than that of dataflow programming in terms of the following three aspects. First, unlike dataflow programming, in our model, we assume that the states of all resources are synchronously changed and the updating process is not observable. Second, in our model, whether the state of a resource is stored or not is determined by analyzing the description of the model, but in dataflow programming, it should be explicitly specified in the program as a kind of self-loop. Last, each data transfer method in a dataflow program is fixed to PUSH-style.

11 Conclusion

We have presented a formal architecture model to abstract some control-independent parts of architecture designs that are not affected by changes of data transfer methods and an architecture level refactoring for changing data transfer methods. We have also developed an algorithm to generate JAX-RS prototypes from the architecture model and selected data transfer methods, and proved the correctness of the algorithm. Furthermore, we have implemented a graph-based refactoring tool and conducted case studies using the tool. The results indicate that the generated prototypes can be used to estimate the impact of the selection on the resulting services performance.

As future work, we want to develop an implementation level refactoring for changing data transfer methods. Extending our architecture model to capture runtime resource creation and deletion is also an important issue.

References

1. Allen, R., Garlan, D.: A formal basis for architectural connection. ACM Trans. Softw. Eng. Method **6**(3), 213–249 (1997)
2. Besta, M., Podstawski, M., Groner, L., Solomonik, E., Hoefler, T.: To push or to pull: on reducing communication and synchronization in graph computations. In: Proceedings of the 26th International Symposium on High-Performance Parallel and Distributed Computing, pp. 93–104 (2017)
3. Fielding, R.T.: Architectural styles and the design of network-based software architectures. Ph.D. thesis, University of California, Irvine (2000)
4. Joseph Goguen, J.T., Wagner, D.: An initial algebra approach to the specification, correctness, and implementation of abstract data types. Technical report, IBM, TJ Watson Research Center (1976)
5. Klein, U., Namjoshi, K.S.: Formalization and automated verification of RESTful behavior. In: Gopalakrishnan, G., Qadeer, S. (eds.) CAV 2011. LNCS, vol. 6806, pp. 541–556. Springer, Heidelberg (2011). https://doi.org/10.1007/978-3-642-22110-1_43
6. Magee, J., Kramer, J., Giannakopoulou, D.: Behaviour analysis of software architectures. In: Proceedings of Working IEEE/International Federation for Information Processing Conference on Software Architecture, pp. 35–49 (1999)
7. Milner, R.: Communication and Concurrency. Prentice-Hall, Hoboken (1990)
8. Pelliccione, P., Inverardi, P., Muccini, H.: Charmy: a framework for designing and verifying architectural specifications. IEEE Trans. Softw. Eng. **35**(3), 325–346 (2009)
9. Sousa, T.B.: Dataflow programming: concept, languages and applications. In: Proceedings of 7th Doctoral Symposium on Informatics Engineering (2012)
10. Sutherland, W.R.: Online graphical specification of computer procedures. Ph.D. thesis, MIT (1966)
11. Veen, A.H.: Dataflow machine architecture. ACM Comput. Surv. **18**, 365–396 (1986)
12. Zhao, Y.: A model of computation with push and pull processing. Technical report UCB/ERL M03/51, EECS Department, University of California, Berkeley (2003). http://www2.eecs.berkeley.edu/Pubs/TechRpts/2003/4192.html

Joint Use of SysML and Reo to Specify and Verify the Compatibility of CPS Components

Perla Tannoury[✉], Samir Chouali, and Ahmed Hammad

University of Bourgogne Franche-Comté, FEMTO-ST Institute - UMR CNRS 6174,
Besançon, France
{perla.tannoury,schouali,ahammad}@femto-st.fr

Abstract. Modeling and verifying the behavior of Cyber-Physical Systems (CPS) with complex interactions is challenging. Traditional languages such as SysML diagrams are not enough to capture CPS coordination. In this paper, we propose a novel approach called SysReo, which extends SysML diagrams (RD, BDD, IBD, SD) with the Reo coordination language. Our main objective is to enhance the interoperability of CPS by providing a more precise representation of system behavior and interaction protocols. To achieve this goal, we extend the SysML sequence diagram (SD) with Reo to create the SysReo SD. Through this integration, we bridge the gap between traditional modeling languages and the coordination demands of CPS. We develop an algorithm to generate Constraint Automata (CA) from SysReo SD, which ensures that CPS components can seamlessly work together. These automata are used in a verification tool that checks formulas expressed in Linear Temporal Logic (LTL). By leveraging LTL and Constraint Automata, we enhance the precision and rigor of CPS verification processes, while guaranteeing that CPS components can seamlessly work together. Furthermore, we apply our approach to a medical CPS case study, illustrating its effectiveness in identifying design flaws early and ensuring system behavior aligns with desired properties.

Keywords: CPS · SysReo · SysML · Reo · Constraint Automata · LTL · Specification · Verification

1 Introduction

In today's technologically advanced world, Cyber-Physical Systems (CPS) have become crucial in a range of applications, such as autonomous vehicles [37], modeling smart city software interactions [38], and healthcare systems [1]. These systems combine the physical and digital worlds, resulting in improved automation, control, and data processing. Despite the importance of CPS, modeling them can be challenging, especially in the healthcare and medical sectors, as it involves integrating various system components, behaviors, and interaction protocols. In

J. Cámara and S.-S. Jongmans (Eds.): FACS 2023, LNCS 14485, pp. 84–102, 2024.
https://doi.org/10.1007/978-3-031-52183-6_5

addition, collaborative efforts between designers, developers, and stakeholders are required to address these challenges. As CPSs continue to become more complex, it is essential to establish an environment that facilitates the modeling process while highlighting their fundamental, structural, and behavioral aspects.

Several languages and formalisms are used to model CPS [17,23,33,34]. In our research, we have opted to use the System Modeling Language (SysML) [36] language due to its ability to model heterogeneous systems that combine hardware and software components. We aim to develop an approach that enables users to easily create specifications using SysML, while taking into account both verification and validation processes. SysML is widely used in industrial applications to model various aspects of a system, including its architecture, behavior, and requirements. However, CPSs are typically composed of various components that interact through various protocols, leading to complex system behaviors. While SysML is a valuable language for describing CPS, it may not be sufficient to formally specify and verify the complex interactions between CPS components. To tackle this problem, we have proposed in our previous work [39,40] a new domain-specific language (DSL) called SysReo that effectively uses the strengths of both semi-formal and formal languages to improve the validation and verification of CPS.

SysReo, is used to overcome the challenges in designing a CPS by clearly expressing its interaction protocols at any design stage. By extending SysML, a semi-formal language, with Reo [5], a formal coordination language, in the SysReo framework, it becomes possible to model the complex components of a CPS in an effective manner. This integration provides a powerful tool for representing component interactions, allowing for more precise and accurate modeling of complex systems. SysReo empowers CPS designers to model all facets of a CPS while explicitly defining the conditions under which data can flow between components. However, it should be noted that SysReo has limitations in formally specifying and verifying the behavioral aspects of CPS. In contrast, the SysML Sequence Diagram (SD) [36] excels in representing component interactions over time in CPS. Nonetheless, SD is semi-formal and lacks direct verification capabilities. On the other hand, Reo offers a formal representation of component coordination and allows for system property analysis. However, stakeholders may find Reo challenging to comprehend due to its complex representation.

In this paper, we first introduce a novel approach called "SysReo Sequence Diagram (SysReo SD)" that enhances the modeling and analysis of CPS. By extending the SysML Sequence Diagram (SD) with Reo notation, we create a "semi-formal-formal" model that captures the behavior and coordination of CPS components using an exogenous protocol. Unlike traditional SysML SD, which focuses on internal system behavior, SysReo SD imposes an external order on the flow of data between components without directly affecting their behavior. The SysReo SD model serves two main purposes: first, it bridges the gap between visual representation and formal modeling, providing a comprehensive view of system behavior and interaction protocols. Second, it enables the formal verification and analysis of the interoperability and correctness of the CPS. To facilitate the verification process, we develop an algorithm that generates Reo Constraint Automata (CA) [15] directly from the SysReo SD model. In the next phase, we use the generated automaton CA as input in the vereofy [12] model checking tool to formally verify Linear Temporal Logic (LTL) [11] properties. This allows us to effectively verify the interoperability of CPS components, confirm compliance with specified requirements, and ensure correct system behavior. To demonstrate the effectiveness of our approach, we conducted a case study involving a smart medical bed, showcasing its potential for real-world applications in the medical CPS domain.

To the best of our knowledge, no previous research has comprehensively explored the extension of SysML Sequence Diagram (SD) with Reo to effectively model the behavior and interaction protocols of CPS. Existing approaches either resulted in verbose and less readable Reo circuits derived from scenario specifications [7,9], or encountered difficulties in establishing correlations between Reo circuits and the original specifications [35]. Our work addresses this gap by enhancing SysML SD with the coordination capabilities of Reo and introducing a novel algorithm that directly generates Constraint Automata (CA) from SysReo SD diagrams. This is followed by a formal verification process to ensure CPS interoperability and validate design correctness.

The paper is structured as follows. Section 2 provides a concise introduction to Reo, Constraint Automata, and SysReo, highlighting their key concepts and features. In Sect. 3, we present the related works. Moving forward, Sect. 4 offers an in-depth case study that showcases the practical application of our SysReo model, focusing on the specification and verification processes involved. Finally, Sect. 5 concludes the paper and briefly discusses future work.

2   Preliminaries

In this section, we give a brief introduction to Reo, constraint automata (CA), and SysReo.

2.1   Reo and Constraint Automata (CA) in a Nutshell

Reo, as described by Arbab in [5], is an external coordination model that prioritizes efficient communication and coordination among different components. It achieves this by using channel-based connectors to establish complex coordinators. However, Reo does not focus on internal activities and communications within individual components. Instead, its main emphasis is on the coordination and interaction between components. The fundamental elements of Reo consist of components, channels, nodes, and ports, working in harmony to enable seamless data exchange and synchronization between these components [5,8,10].

The formal semantics of Reo are rigorously captured through the use of Constraint Automata (CA) [15]. CA provides a systematic representation of interactions among anonymous components, describing behavior and data flow in coordination models. This formalism involves labeling transitions with sets of ports that are triggered simultaneously, complemented by data constraints applied to these ports. As a result, constraint automata offers a powerful means to precisely specify and analyze component interactions within the Reo coordination model.

Definition 1. *Constraint Automata (CA):* A constraint automaton B = (S,S_0,N,δ) is composed of:

- S: set of states (or locations).
- S_0: initial state where $S_0 \in S$.
- N: set of port names.
- δ: transition relation $\delta \subseteq S \times 2^N \times DC \times S$, where DC is the set of Data Constraints (DC) over a finite data domain Data.

An example of a constraint automata B is illustrated in Fig. 6 **step2**, where B = $(\{S_0, S_1\}, S_0, \{(A,V), (B,W)\}, \{(S_0, (A,V), [A] = |V|, S_1), (S_1, (B,W), [B] = |W|, S_0)\})$.

2.2 SysReo

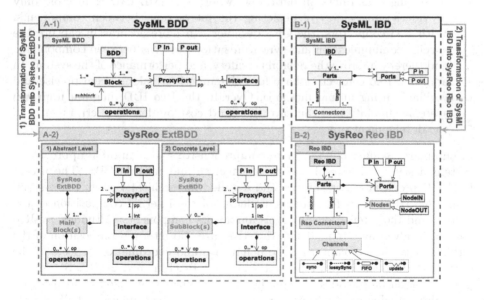

Fig. 1. From SysML diagrams to SysReo diagrams.

SysReo was introduced in [39, 40] as a powerful modeling language that combines the strengths of both SysML and Reo to provide a more comprehensive and flexible approach to model a CPS. One of the main advantages of SysReo is its ability to model all facets of CPS, from requirements to architecture and interaction protocols, which is essential for designing complex heterogeneous systems. Figure 1 represents the process of transforming the SysML Block Definition

Diagram (BDD) and Internal Block Diagram (IBD) into SysReo Extended Block
Definition Diagram (ExtBDD) and Reo Internal Block Diagram (Reo IBD). The
transformation consists of two parts:

(A) The SysML BDD (Fig. 1 A-1) is transformed into SysReo ExtBDD (Fig. 1 A-
    2) by using SysML BDD meta-models and dividing the CPS hierarchy into
    two levels. The first level presents the abstract model of the CPS, where the
    primary components of the system are modeled as main blocks. The second
    level presents the concrete components of the CPS, where they are modeled
    as sub-blocks. Overall, the traditional SysML BDD offers a high-level view
    of the system being designed, whereas the Extended BDD offers a more in-
    depth view of the system structure. Using multiple levels of abstraction and
    additional information in the ExtBDD can help to control the complexity
    of the system being designed and ensure that the system design meets the
    intended requirements.
(B) The SysML IBD (Fig. 1 B-1) is converted into SysReo Reo IBD (Fig. 1 B-2)
    by using SysML IBD meta-models and replacing the IBD connectors with
    "Reo connectors". This replacement allows for more explicit representation
    of the internal composition of components and their interaction protocols
    by setting constraints on data flow, whereas SysML IBD connectors only
    provide a generic way of depicting the interactions between components.
    With Reo connectors, the CPS designer can more accurately capture the
    specific communication and synchronization patterns between components,
    and thereby improve the reliability, safety, and performance of the system. In
    addition, Reo connectors have formal semantics, making them precise and
    verifiable using formal methods. Overall, the Reo IBD diagram improves
    system reliability and can save time and cost by detecting errors early in
    the development process.

In summary, SysReo offers a more comprehensive and adaptable approach to
system design, which results in more effective and reliable CPS. The use of
ExtBDD provides a more precise and detailed hierarchical view of the system,
while Reo IBD allows for explicit modeling of the internal composition of the
system and the interaction protocols among its components. Although SysReo
has many advantages in modeling the structure and internal composition of the
CPS, it falls short of capturing the behavioral and coordination aspects of the
CPS. To bridge this gap, we present a new approach named SysReo SD in this
paper. Section 4.3 provides a comprehensive overview of SysReo SD, focusing on
its ability to effectively address the behavioral and coordination challenges in
CPS.

3   Related Works

CPSs are networks of different embedded systems connected in a physical environment, making their specification and formal verification difficult due to their
complex and large-scale computing infrastructure.

Previous works in [3, 22, 26, 32] proposed different approaches to model different aspects of CPS with SysML. To formally verify critical safety systems, the authors in [18, 19] proposed to extend the SysML sequence diagram (SD) and automate consistency verification using the Clock Constraint Specification Language (CCSL) [4]. Additionally, an approach was proposed [16] to transform SD into interface automata (IA) to verify the compatibility and consistency of components modeled with SysML. However, these works take an endogenous approach to coordination and neglect the coordination of message exchanges, leading to communication problems within CPSs. To address this problem, Reo is proposed as a coordination language to fill the interfacing gaps and enhance the modeling and coordination of CPSs.

Various researchers explored different approaches to model CPS using Reo and have demonstrated its effectiveness through formal analysis techniques. These include co-algebraic semantics [10], operational semantics [15] using constraint automata [15] and timed constraint automata [6, 30], coloring semantics [21], and converting Reo models to other formal models such as Alloy [28] and mCRL2 [31] to leverage existing verification tools. Despite its advantages, Reo remains complex for stakeholders to comprehend due to its lack of semi-formal representation.

To our knowledge, there is no prior research that explores the extension of SysML Sequence Diagrams (SD) with Reo to effectively model behavior and interaction protocols in CPS. Previous approaches resulted in complex and less comprehensible Reo circuits derived from scenario specifications [7, 9], or faced challenges in establishing connections between the Reo circuits and the original specifications [35]. In our work, we bridge this gap by enhancing SysML SD with Reo's coordination capabilities and introducing a novel algorithm that directly generates Constraint Automata (CA) from SysReo SD diagrams. This is followed by a formal verification process to ensure interoperability of the CPS and validate the accuracy of the design, addressing the limitations of existing approaches.

4   Case Study: Smart Medical Bed (SMB)

In this section, we present our case study of the Smart Medical Bed (SMB) system. First, we begin by briefly introducing the SMB system. Then, we gather information about the SMB system and analyze it using our SysReo models. This process involves specifying the system's requirements, designing its structure and internal composition, and modeling the system's behavior and interaction protocol. Finally, we move on to the verification phase, where we rigorously verify the correctness of our SysReo models.

4.1   SMB Overview

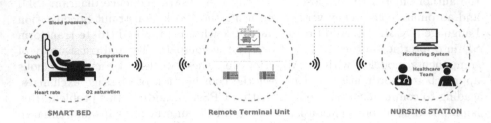

Fig. 2. Smart medical bed architecture.

A Smart Medical bed (SMB) is equipped with various sensors and monitoring devices that collect data on vital signs of the patient, such as heart rate, blood pressure, temperature, oxygen saturation, and other key indicators of their health. These data are then transmitted to a Remote Terminal Unit (RTU), where it can be stored and analyzed by healthcare providers as we can see in Fig. 2. RTU plays a key role in connecting, controlling, analyzing, and communicating data from the smart bed to the nursing station, providing valuable information to the healthcare team and ultimately improving patient care. Continuous monitoring helps identify potential health issues early, leading to informed decision-making and efficient care delivery. The SMB infrastructure comprises the following components: (1) the Smart Bed (SB), where the patient resides, (2) the Remote Terminal Unit (RTU), where the collected data are stored, processed and analyzed, and (3) the Nursing Station (NS), where the monitoring system and the healthcare team are located. In this article, we focus on modeling and verifying the requirement, structure, behavior, and interaction protocol between the Smart Bed (SB) and the RTU in the SMB system. Using our model-driven approach, SysReo, we can analyze the system's requirements, model the architecture and internal structure of the SMB. To handle complexity, we introduce SysReo SD, an extension for modeling complex component behavior and interaction protocols.

4.2   Modeling SMB with SysReo

In this section, we focus on our modeling approach. As shown in Fig. 3, the process is broken down into two phases. During the first phase, the CPS designer begins by collecting requirements about the system and analyzing it. Using our SysReo model, the designer then specifies the system's needs, which results in three main diagrams: (1) The requirement diagram that models the functional and non-functional needs of the system. (2.1) The ExtBDD diagram that represents the hierarchical structure of the system as blocks, followed by (2.2) the Reo IBD diagram that is used to model the system's internal structure and interaction protocols. (3) The SysReo sequence diagram (SysReo SD) used to model

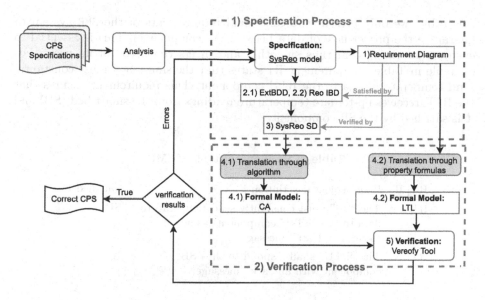

Fig. 3. Specification and verification process.

the behavior and coordination of CPS components. Finally, to satisfy our predefined requirements, we link them to the Reo IBD diagram, and to verify them,
we establish a link to SysReo SD.

In the second phase, we focus on the verification process. To verify the interoperability of CPS components, SysReo SD is translated into CA through our
algorithm that directly generates CA from SysReo SD. Then the requirements to
Reo IBD are formally defined through property formulas such Linear Temporal
Logic (LTL) [11] for verification. This step is crucial to achieve the objective
of our modeling approach: creating a precise representation of the CPS system's needs, behavior, and coordination. By generating CA and translating the
requirements into LTL formulas, the CPS designer can ensure both the interoperability of the CPS components and the accurate representation of requirements.
Furthermore, the formal verification of these requirements can be seamlessly
conducted using specialized verification tools like vereofy [12,14]. This helps to
guarantee that the CPS system will function as intended and meets the designer's
requirements.

After evaluating the verification results, the CPS designer checks if there are
any specification errors. If so, the process returns to the SysReo model specification phase until a correct CPS model is obtained.

4.3   Specification Process: SysReo Models

Requirement. The design process for any system is crucial for ensuring its
functionality and usability. The first step in this process is to identify the specific
needs of the system, as outlined in the requirements table such as Table 1. In

this table, we only present two functional requirements of the SMB system to guarantee the proper flow of data between its components. This is essential to ensure that the system runs smoothly and meets the needs of its users. For example in Table 1, requirement R1 states that the smart bed must constantly send temperature data to the RTU component. This requirement ensures that the RTU receives up-to-date temperature readings from the smart bed (SB) and it is satisfied by the SB component.

Table 1. Requirement table of SMB.

Req ID	Requirement description	Satisfied by
R1	The "SB" must constantly send temperature data to the "RTU" component using "sendTempData" message	SB
R2	The "RTU" shall respond to the "SB" component with an "ack" message.	RTU

Based on the requirements of Table 1, we can identify the main components of the SMB system that are the Smart Bed (SB) and the Remote Terminal Unit (RTU). In the next section (Sect. 4.3), these components will be used to create a hierarchical view of the SMB system, to better understand the relationships and overall functioning of the system.

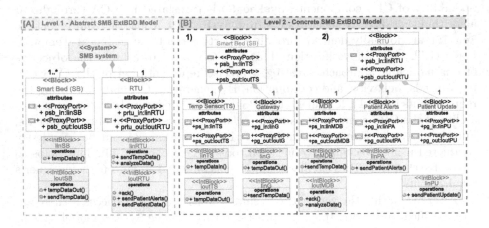

Fig. 4. The ExtBDD model of the SMB system.

ExtBDD. The Extended Block Definition Diagram is used to model the hierarchical view of the SMB system where each component is modeled as a block.

The block defines a component by describing its internal operations as private operations of the block and also its required and offered services as follows. Each block is decorated by two proxy ports: 1) Input port that provides information on the services that are available. These services are listed in an interface block which specifies the type of the port. 2) Output port that describes the required services in a similar manner. Figure 4[A] represents the abstract part of the system by only showing the main components. It consists of the block named "SMB" that represents the system as a whole. It is decomposed into two sub-blocks: Smart Bed (SB) and the Remote Terminal Unit (RTU), in which it is linked to them by the composition relationship. Figure 4[B] shows the concrete level of the SMB system. It depicts the sub-components that make up each main component. As an example, the smart bed component is broken down into two blocks, a Temperature Sensor and a Gateway. The main function of the Temperature Sensor is to continuously measure, record, gather and transmit the measured data to the Gateway.

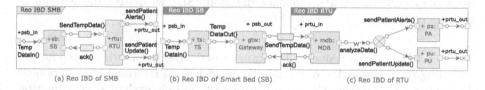

(a) Reo IBD of SMB   (b) Reo IBD of Smart Bed (SB)   (c) Reo IBD of RTU

Fig. 5. Reo IBD of SMB, SB, and RTU.

Reo IBD. The Reo Internal Block Diagram is a combination of the coordination language Reo and the SysML Internal Block Diagram (IBD) [24] where the IBD's connectors are translated into Reo circuits. Reo IBD is used to characterize the internal components and structure of a system block, including its properties, parts, connections, and interaction protocols. It explicitly specifies the rules and conditions of the data that can be transferred from the component's input to its output through channels. In Fig. 5(a), two different channels are used to model the interaction protocols among the following main components: the Smart Bed (SB) and Remote Terminal Unit (RTU).

1. FIFO channel: for example, two FIFO channels "sendTempData()" and "ack()" are used to model the asynchronous communication between the two components "SB" and "RTU". The FIFO channel helps to efficiently manage memory in a real-time system by processing the oldest data first, thus preventing loss of information.
2. Sync channel: used to model the synchronization properties among components.

Figure 5(b) models the internal structure of the smart bed. First of all, the "Temperature Sensor (TS)" sends the vital signs data of the patients to the

"Gateway" component via the "TempDataOut()" flow port using a sync channel. Once received, the "Gateway" component sends the data, which has been pre-processed, to the RTU through the "Medical Data Base (MDB)" component using a FIFO channel "SendTempData()". Once it is received, the MBD replies to the smart bed component with an "ack()" using a fifo channel. Then in Fig. 5(c), the "MDB" component analyzes the data and sends "analyzeData()" to xrouter component (⊗) via a filter channel where it models the routing replication of data to "Patient Alerts (PA)" or to "Patient Update (PU)". Once data enters the xrouter component, it is sent either to "PA" or to "PU" but never to both.

Reo IBD has many advantages when it comes to modeling a CPS. Such as effectively modeling component connections, characterizing message flow, satisfying predefined requirements, and enabling design flexibility while simplifying documentation with its graphical representation. However, it is limited to modeling temporal behaviors and formally verifying predefined requirements, which are crucial aspects in correctly modeling a CPS. Therefore, in the next section, we will extend the SysML sequence diagram (SD) with Reo and explore its benefits on CPSs.

SysReo SD. SysReo Sequence Diagram extends SysML SD with Reo notation, including the introduction of the Reo Sequencer as an intermediary component. This integration enriches the representation of message exchange, coordination, and synchronization in a unified manner. In contrast to conventional SysML sequence diagrams, SysReo SD facilitates the explicit specification of protocols, eliminating the need for manual implementation of locks and buffers. This approach enhances accuracy, efficiency, and mitigates errors typically associated with manual synchronization mechanisms. As a result, SysReo SD offers a robust and comprehensive approach for specifying protocols in system behavior, particularly advantageous for capturing communication flow and coordination patterns within a single diagram. Figure 6(A) presents a simple example of a SysML SD,

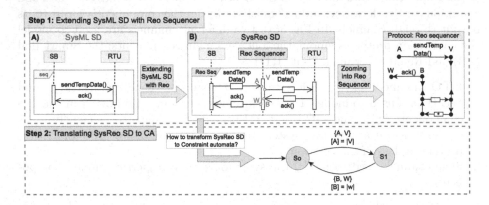

Fig. 6. Extending SysML SD with Reo then translating SysReo SD to CA.

demonstrating the behavior and message exchange between the two components of SMB, smart bed (SB) and RTU. This coordination follows an endogenous approach where protocols are implicitly expressed within the embedded code fragments. However, modifying these protocols can be challenging and may require extensive changes to multiple software components, potentially impacting previously validated properties.

On the other hand, exogenous methods like Reo offer a more explicit and modular approach to define protocols. In the given scenario Fig. 6(A), where components SB and RTU need to exchange messages, an endogenous approach would involve directly implementing the exchange within their respective code. However, in Fig. 6(B) illustrates a different strategy, where a separate component called the "Reo sequencer" explicitly defines the message exchange protocol between SB and RTU. For example, SB sends "sendTempData()" to RTU using reo ports {A, V} (depicted as red circles), and the Reo sequencer coordinates this message exchange. This decoupling allows for easier protocol modifications without affecting the implementation of SB and RTU. Employing Reo connectors in exogenous approaches provides more flexibility and simplifies the specification and adjustment of complex protocols in CPS.

The next step involves converting SysReo SD into constraint automata. To achieve this, we will first present a formal definition of SysReo SD. Subsequently, in Sect. 4.4, we will outline an algorithm that facilitates the automatic generation of constraint automata from SysReo SD.

Formal Definitions of SysReo SD. This section presents the formal definition of SysReo message and SysReo SD.

Definition 2 *(SysReoMes).* A SysReo message is a tuple SysReoMes $= (comp_s,$ action, $comp_f$, P, Σ) where:

- $comp_s$: is the source component of the SysReo message.
- *action*: is the called method.
- $comp_f$: is the target component of the SysReo message.
- *P*: is the set of Reo ports.
- Σ = input, output is the set of synchronization constraints, specifying the allowed input/output actions on Reo ports.

For example in Fig. 6(B) a SysReo message between the component SB and RTU can be defined as SysReoMes = (SB, sendTempData, RTU, {A, V}, {[A],|V|}).

Definition 3 *(SysReo SD).* A SysReo SD is defined as tuple B = (IM, SysReoMes, ReoSeq, ReoLoop, ReoAlt) is composed of:

- *IM*: is the initial message.
- *SysReoMes*: the set of messages in SysReo SD.
- *ReoSeq* $= (ReoSeq_1, ...,ReoSeq_i, ... ,ReoSeq_n)$ is the list of reo sequencer combined fragments. $ReoSeq_i = (obj_1, ... , obj_i, ...,obj_n)$, obj_i is message or a fragment and card$(ReoSeq_i) \geq 2$.

- $ReoLoop = (ReoLoop_1, ..., ReoLoop_i, ..., ReoLoop_n)$ is the list of reo loop combined fragments. $ReoLoop_i = (obj_1, ..., obj_i, ..., obj_n)$, obj_i is message or a fragment and card($ReoLoop_i$) ≥ 1.
- $ReoAlt = (ReoAlt_1, ..., ReoAlt_i, ..., ReoAlt_n)$ is the list of reo alternator combined fragments. $ReoAlt_i = (obj_1, ..., obj_i, ..., obj_n)$, obj_i is message or a fragment and card($ReoAlt_i$) ≥ 2.

In this paper, we focus on the Reo sequencer from the SysReo SD fragments (ReoSeq, ReoLoop, and ReoAlt) mentioned above. Our main objective is to translate SysReoSD = $(IM, SysReoMes, ReoSeq)$ into reo constraint automata using our algorithm. This automaton will serve as input to vereofy tool where we can formally verify the correctness and interoperability of SMB system through LTL properties.

4.4   Verification Process: CA, Vereofy, LTL

This section outlines the verification process of SysReo. First, we start by proposing an algorithm to construct CA from SysReo SD diagram. Then, the resulting automaton is used as input for the vereofy tool to assess the accuracy and interoperability of the SMB system through the application of LTL properties.

Algorithm. SysReo SD is a visual representation of the SMB system, which consists of components and their interactions. However, to analyze and verify the system formally, it is necessary to transform this visual representation into a more structured and formal representation. Therefore through our algorithm, we provide a systematic and automated approach to convert the SysReo SD into constraint automata, which are well-suited for formal analysis.

Algorithm 1 is developed to transform a SysReo SD into a Constraint Automaton (CA), using a SysReo message and fragment list as inputs. The resulting CA comprises a set of states (S), an initial state (S_0), a set of port names (N), and a set of transitions (δ). The algorithm begins by initializing the set of states, checking for emptiness, and creating a new state as the initial state (cf. Algorithm 1, lines 2–7). It then iterates through each object in the list. If the object is part of a Reo sequencer fragment, the algorithm recursively calls itself with the corresponding sub-list representing that fragment (cf. Algorithm 1, lines 8–13). When encountering a SysReo message object ($SysReoMes$), the algorithm creates a new state s' if the message is not the last object in the list (cf. Algorithm 1, lines 14–18). If it is the last object, s' is designated as the initial state s_0. The algorithm then populates the sets S and N in the CA with the newly created state and the ports of the $SysReoMes$, respectively. A new transition is created from the previously added state s to s', incorporating the port name and constraints from the $SysReoMes$, and it is subsequently added to the set of transitions δ. This process continues until there are no remaining objects in the list to be processed (cf. Algorithm 1 line 19→30).

The algorithm, SysReoSDtoCA, has a linear complexity dependent on the size of the objects set l in the SysReoSD diagram specification. When we apply

---

Algorithm 1: Mapping SysReoSD to CA algorithm

---

1 **Function** SysReoSDtoCA(SysReoSD, l, CA)
  Input: SysReoSD = (IM, SysReoMes, ReoSeq);
  l: a list of objects in SysReoSD;
  $l = (obj_1, ..., obj_i, ..., obj_n)$,
  obj_i is a message or a fragment in SysReoSD.
  Output: CA=(S,S_0, N, δ)

2 **Begin**
3 **if** $S = \emptyset$ **then**
4     $s = $ createNewState()
5     $S = S \cup \{s\}$ // add s to S (the set of states in CA)
6     $S_0 = s$ // set the initial state S_0 to s
7 **end**
8 **while** $(l \neq \emptyset)$ **do**
9   Let obj the first element in l
10   **if** $obj \in ReoSeq$ **then**
11   Let l' be a list of objects composing the Reo sequencer fragment
12   SysReoSDtoCA(SysReoSD, l', CA)
13   **end**
14   **else**
15   **if** $obj \in SysReoMes$ **then**
16   **if** $card(l) > 1$ **then**
17   s'= createNewState()
18   **end**
19   **else**
20   s'= S_0
21   **end**
22   S=S $\cup \{s, s'\}$ // the set of states
23   N=N $\cup$ obj.P // the set of port names
24             $\delta = \delta \cup$ (s, obj.P, obj.Σ, s') // the set of transitions
25   s=s' // initial state
26   **end**
27   **end**
28   l=l'- {obj}
29 **end**
30 **End SysReoSDtoCA**

---

this algorithm to the formal model of the SysReoSD diagram example, as presented in Fig. 6B, we obtain the constraint automata CA = (S,S_0,N,δ) described in Fig. 6 **step2**. The CA is characterized by the following components:

1. Set of states: S=$\{S_0, S\}$.
2. Initial state: $S_0 = S_0$.
3. Set of port names: N= $\{(A,V), (B,W)\}$.
4. Set of transitions: $\delta=$ S x $2^{A,V}$ x DC x S_0 where DC = [A], $|V|$, and the second $\delta=$ S_0 x $2^{B,W}$ x DC x S where DC = [B], $|W|$.

Upon analyzing the resulting constraint automaton, we can observe the system's behavior when component SB sends the SysReoMes "sendTempData" to component RTU. This communication is coordinated by the Reo sequencer and specified by the transition labeled with reo ports {A, V} and input/output action interfaces [A], $|V|$. As a response, component RTU sends the SysReoMes "ack" back to component SB through the reo sequencer ports {B, W} and input/output action interfaces [B], $|W|$.

Vereofy Tool. Vereofy [12, 14], developed at the University of Dresden, is a powerful model checking tool specifically designed for analyzing and verifying Reo connectors. It supports two input languages: the Reo Scripting Language (RSL) for specifying coordination protocols, and the Constraint Automata Reactive Module Language (CARML), a textual representation of constraint automata used to define component behavior. With vereofy, one can verify temporal properties expressed in LTL [11] and CTL-like logics [29]. Distinguishing itself from other model checkers [2, 20, 25], vereofy places a primary focus on verifying coordination aspects, communication, and interactions at the behavioral interface level. It employs a symbolic representation based on binary decision diagrams (BDDs) to facilitate efficient verification algorithms. For a detailed understanding of the modeling languages and verification techniques employed by vereofy, refer to [12, 13].

Next, we present the CARML code corresponding to the generated Constraint Automaton (CA) from SysReoSD. The code module provided below specifically defines a sysreoCA. Initially, the sysreoCA is in an empty state denoted as S_0. When the sysreoCA is not in a full state S_1, and a data value is written to its input port A, the data is stored in the 'sendTempData' variable, and the internal state changes from S_0 to S_1. Another component reading data from the output port V resets the internal state back to S_0. The data domain has been locally set to the integer range (0,1). Although it is possible to set the data domain to any other available datatype, it can only be done once across the included files or as a runtime argument. The data flow within the system may depend on the value of a variable of type 'Data', as illustrated in the second transition, where the 'sendTempData' stored is written to the output port V, represented as: $\#V == sendTempData$.

```
#Vereofy CARML code:
TYPE Data = int(0,1);

MODULE sysreoCA{
// initializing the I/O ports {A,V} and { B,W}
  in: A;
  out: V;
  in: B;
  out: W;

//defining the set of states that are S0 and S1
```

```
var : enum {S0, S1} state := S0;

//defining the messages that should be exchanged
var : Data sendTempData := 0;
var : Data ack      := 0;

//drawing the transitions from So to S1 and from S1 to S0
state = = S0  -[ {A} ]-> state:=S1 & sendTempData:=#A;
state = = S1  -[{V} & #V = = sendTempData ]-> state:=S0;

state = = S0  -[ {B} ]-> state:=S1 & ack:=#B;
state = = S1  -[{W} & #W = = ack ]-> state:=S0;
}
```

Verification of LTL Properties. LTL-based model checking rigorously verifies system properties, boosts confidence in correctness and reliability, and identifies design flaws for improvements. To verify the correctness of the specified protocol and the interoperability between the smart bed and RTU, we propose to check the following LTL properties in vereofy tool:

```
p1:LTL<<G(("{A}" & "#A==1")->X("state==S1"& "sendTempData==1"))>>
/*PASSED*/


p2:LTL<< G(("{W}" & "#W==1")-> X("state==S0" & "ack==1"))>>
/*PASSED*/
```

The evaluated LTL property in "p1" ensures that when port A is active once, the subsequent state must satisfy two conditions: the state variable should be S_1 (indicating a transition from S_0 to S_1) and the 'sendTempData' variable should be 1 (indicating successful message transmission). The result, "PASSED", confirms that this property holds for all execution traces. The same thing applies for the evaluated LTL property in "p2" where it indicates the successful transmission of the "ack" message.

5 Conclusion

Our paper introduces a novel diagram called "SysReo SD" that enhances CPS modeling and analysis. By extending SysML with Reo, we create a powerful "semi-formal-formal" model that effectively captures the behavior and coordination of CPS components using an exogenous protocol. This allows us to ensure CPS interoperability, meet specific design requirements, and validate the correctness of the system behavior. Furthermore, we illustrate the applicability of our approach through a case study in the medical CPS domain, showcasing the potential benefits of employing SysReo.

Looking ahead, we are planning to explore the application of SysReo in Digital Twins (DT) [27] where we can accurately capture the interactions and behaviors of the components of a physical system in a virtual environment. With the use of SysReo models, we can continuously monitor and optimize the performance of the digital twin.

References

1. Al-Jaroodi, J., Mohamed, N., Abukhousa, E.: Health 4.0: on the way to realizing the healthcare of the future. IEEE Access **8**, 211189–211210 (2020)
2. Alur, R., et al.: jMocha: a model checking tool that exploits design structure. In: Proceedings of the 23rd International Conference on Software Engineering. ICSE 2001, pp. 835–836. IEEE (2001)
3. Amálio, N., Payne, R., Cavalcanti, A., Brosse, E.: Foundations of the SysML profile for CPS modelling. Deliverable D2. 1a, version 1 (2015)
4. André, C.: Syntax and semantics of the clock constraint specification language (CCSL). Ph.D. thesis, INRIA (2009)
5. Arbab, F.: Reo: a channel-based coordination model for component composition. Math. Struct. Comput. Sci. **14**(3), 329–366 (2004). https://doi.org/10.1017/S0960129504004153
6. Arbab, F., Baier, C., de Boer, F., Rutten, J.: Models and temporal logical specifications for timed component connectors. Softw. Syst. Model. **6**, 59–82 (2007)
7. Arbab, F., Baier, C., de Boer, F., Rutten, J., Sirjani, M.: Synthesis of Reo circuits for implementation of component-connector automata specifications. In: Jacquet, J.-M., Picco, G.P. (eds.) COORDINATION 2005. LNCS, vol. 3454, pp. 236–251. Springer, Heidelberg (2005). https://doi.org/10.1007/11417019_16
8. Arbab, F., Baier, C., Rutten, J., Sirjani, M.: Modeling component connectors in Reo by constraint automata. Electron. Notes Theoret. Comput. Sci. **97**, 25–46 (2004)
9. Arbab, F., Meng, S.: Synthesis of connectors from scenario-based interaction specifications. In: Chaudron, M.R.V., Szyperski, C., Reussner, R. (eds.) CBSE 2008. LNCS, vol. 5282, pp. 114–129. Springer, Heidelberg (2008). https://doi.org/10.1007/978-3-540-87891-9_8
10. Arbab, F., Rutten, J.J.M.M.: A coinductive calculus of component connectors. In: Wirsing, M., Pattinson, D., Hennicker, R. (eds.) WADT 2002. LNCS, vol. 2755, pp. 34–55. Springer, Heidelberg (2003). https://doi.org/10.1007/978-3-540-40020-2_2
11. Babenyshev, S., Rybakov, V.: Linear temporal logic LTL: basis for admissible rules. J. Log. Comput. **21**(2), 157–177 (2011)
12. Baier, C., Blechmann, T., Klein, J., Klüppelholz, S.: Formal verification for components and connectors. In: de Boer, F.S., Bonsangue, M.M., Madelaine, E. (eds.) FMCO 2008. LNCS, vol. 5751, pp. 82–101. Springer, Heidelberg (2009). https://doi.org/10.1007/978-3-642-04167-9_5
13. Baier, C., Blechmann, T., Klein, J., Klüppelholz, S.: A uniform framework for modeling and verifying components and connectors. In: Field, J., Vasconcelos, V.T. (eds.) COORDINATION 2009. LNCS, vol. 5521, pp. 247–267. Springer, Heidelberg (2009). https://doi.org/10.1007/978-3-642-02053-7_13

14. Baier, C., Blechmann, T., Klein, J., Klüppelholz, S., Leister, W.: Design and verification of systems with exogenous coordination using vereofy. In: Margaria, T., Steffen, B. (eds.) ISoLA 2010. LNCS, vol. 6416, pp. 97–111. Springer, Heidelberg (2010). https://doi.org/10.1007/978-3-642-16561-0_15

15. Baier, C., Sirjani, M., Arbab, F., Rutten, J.: Modeling component connectors in Reo by constraint automata. Sci. Comput. Program. **61**(2), 75–113 (2006)

16. Bouaziz, H., Chouali, S., Hammad, A., Mountassir, H.: SysML model-driven approach to verify blocks compatibility. Int. J. Comput. Aided Eng. Technol. **11**(2), 206–231 (2019)

17. Bouskela, D., et al.: Formal requirements modeling for cyber-physical systems engineering: an integrated solution based on form-l and modelica. Requirements Eng. **27**(1), 1–30 (2022)

18. Chen, X., Liu, Q., Mallet, F., Li, Q., Cai, S., Jin, Z.: Formally verifying consistency of sequence diagrams for safety critical systems. Sci. Comput. Program. **216**, 102777 (2022)

19. Chen, X., Mallet, F., Liu, X.: Formally verifying sequence diagrams for safety critical systems. In: 2020 International Symposium on Theoretical Aspects of Software Engineering (TASE), pp. 217–224. IEEE (2020)

20. Cimatti, A., Clarke, E., Giunchiglia, F., Roveri, M.: NuSMV: a new symbolic model verifier. In: Halbwachs, N., Peled, D. (eds.) CAV 1999. LNCS, vol. 1633, pp. 495–499. Springer, Heidelberg (1999). https://doi.org/10.1007/3-540-48683-6_44

21. Clarke, D., Costa, D., Arbab, F.: Connector colouring i: synchronisation and context dependency. Sci. Comput. Program. **66**(3), 205–225 (2007)

22. DeTommasi, G., Vitelli, R., Boncagni, L., Neto, A.C.: Modeling of MARTe-based real-time applications with sysML. IEEE Trans. Industr. Inf. **9**(4), 2407–2415 (2012)

23. Genius, D., Apvrille, L.: Hierarchical design of cyber-physical systems. In: Modelsward (2023)

24. 0 Hause, M., et al.: The SysML modelling language. In: Fifteenth European Systems Engineering Conference, vol. 9, pp. 1–12 (2006)

25. Holzmann, G.J.: The model checker spin. IEEE Trans. Software Eng. **23**(5), 279–295 (1997)

26. Huang, P., Jiang, K., Guan, C., Du, D.: Towards modeling cyber-physical systems with SysML/MARTE/pCCSL. In: 2018 IEEE 42nd Annual Computer Software and Applications Conference (COMPSAC), vol. 1, pp. 264–269. IEEE (2018)

27. Juarez, M.G., Botti, V.J., Giret, A.S.: Digital twins: review and challenges. J. Comput. Inf. Sci. Eng. **21**(3), 030802 (2021)

28. Khosravi, R., Sirjani, M., Asoudeh, N., Sahebi, S., Iravanchi, H.: Modeling and analysis of Reo connectors using alloy. In: Lea, D., Zavattaro, G. (eds.) COORDINATION 2008. LNCS, vol. 5052, pp. 169–183. Springer, Heidelberg (2008). https://doi.org/10.1007/978-3-540-68265-3_11

29. Kokash, N., Arbab, F.: Formal design and verification of long-running transactions with extensible coordination tools. IEEE Trans. Serv. Comput. **6**(2), 186–200 (2011)

30. Kokash, N., Jaghoori, M.M., Arbab, F.: From timed Reo networks to networks of timed automata. Electron. Notes Theoret. Comput. Sci. **295**, 11–29 (2013)

31. Kokash, N., Krause, C., De Vink, E.: Reo+ mCRL2: a framework for model-checking dataflow in service compositions. Formal Aspects Comput. **24**(2), 187–216 (2012)

32. Larsen, P.G., et al.: Integrated tool chain for model-based design of cyber-physical systems: the into-cps project. In: 2016 2nd International Workshop on Modelling, Analysis, and Control of Complex CPS (CPS Data), pp. 1–6. IEEE (2016)

33. Lin, J., Sedigh, S., Miller, A.: Modeling cyber-physical systems with semantic agents. In: 2010 IEEE 34th Annual Computer Software and Applications Conference Workshops, pp. 13–18. IEEE (2010)

34. Mallet, F.: MARTE/CCSL for modeling cyber-physical systems. In: Drechsler, R., Kühne, U. (eds.) Formal Modeling and Verification of Cyber-Physical Systems, pp. 26–49. Springer, Wiesbaden (2015). https://doi.org/10.1007/978-3-658-09994-7_2

35. Meng, S., Arbab, F., Baier, C.: Synthesis of Reo circuits from scenario-based interaction specifications. Sci. Comput. Program. **76**(8), 651–680 (2011)

36. OMG: OMG System Modeling Language. https://www.omg.org/spec/SysML/. Accessed 10 Feb 2023

37. Panahi, V., Kargahi, M., Faghih, F.: Control performance analysis of automotive cyber-physical systems: a study on efficient formal verification. ACM Trans. Cyber-Phys. Syst. (2022)

38. Pundir, A., Singh, S., Kumar, M., Bafila, A., Saxena, G.J.: Cyber-physical systems enabled transport networks in smart cities: challenges and enabling technologies of the new mobility era. IEEE Access **10**, 16350–16364 (2022)

39. Tannoury, P.: An Incremental Model-Based Design Methodology to Develop CPS with SysML/OCL/Reo. In: Journées du GDR GPL. Vannes, France (2022). https://hal.science/hal-03893454

40. Tannoury, P., Chouali, S., Hammad, A.: Model driven approach to design an automotive CPS with SysReo language. In: Proceedings of the 20th ACM International Symposium on Mobility Management and Wireless Access, pp. 97–104 (2022)

From Reversible Computation
to Checkpoint-Based Rollback Recovery
for Message-Passing Concurrent
Programs

Germán Vidal[✉]

VRAIN, Universitat Politècnica de València, Valencia, Spain
gvidal@dsic.upv.es

Abstract. The reliability of concurrent and distributed systems often depends on some well-known techniques for fault tolerance. One such technique is based on checkpointing and rollback recovery. Checkpointing involves processes to take snapshots of their current states regularly, so that a rollback recovery strategy is able to bring the system back to a previous consistent state whenever a failure occurs. In this paper, we consider a message-passing concurrent programming language and propose a novel rollback recovery strategy that is based on some explicit checkpointing operators and the use of a (partially) reversible semantics for rolling back the system.

Keywords: reversible computation · message-passing · concurrency · rollback recovery · checkpointing

1 Introduction

The reliability of concurrent and distributed systems often depends on some well-known techniques for fault tolerance. In this context, a popular approach is based on *checkpointing* and *rollback recovery* (see, e.g., the survey by Elnozahy *et al.* [3]). Checkpointing requires each process to take a snapshot of its state at specific points in time. This state is stored in some stable memory so that it can be accessed in case of a failure. Then, if an unexpected error happens, a recovery strategy is responsible of rolling back the necessary processes to a previous checkpoint so that we recover a consistent state of the complete system and normal execution can be safely resumed.

In this paper, we consider the definition of a rollback recovery strategy based on three explicit operators: check, commit, and rollback. The first operator, check, is used to define a checkpoint, thus saving the current state of a process. The

This work has been partially supported by grant PID2019-104735RB-C41 funded by MCIN/AEI/ 10.13039/501100011033, by French ANR project DCore ANR-18-CE25-0007, and by *Generalitat Valenciana* under grant CIPROM/2022/6 (FassLow).

J. Cámara and S.-S. Jongmans (Eds.): FACS 2023, LNCS 14485, pp. 103–123, 2024.
https://doi.org/10.1007/978-3-031-52183-6_6

checkpoint is assigned a fresh identifier τ. Then, one can either *commit* the computation performed so far (up to the checkpoint), commit(τ), or roll back to the state immediately before the checkpoint, rollback(τ).

There are several possible uses for these operators. For example, the functional and concurrent language Erlang [5] includes the usual try_catch expression, which in its simplest form is as follows: "try e catch _ : _ $\rightarrow e'$ end." Here, if the evaluation of expression e terminates with some value, then try_catch reduces to this value. Otherwise, if an exception is raised (no matter the exception in this example since "_ : _" catches all of them), the execution jumps to the catch statement and e' is evaluated instead. However, the actions performed during the incomplete evaluation of e are not undone, which may give rise to an inconsistent state of the system. Using the above operators, we could write down a safer version of the try_catch expression above as follows:

$$\textbf{try } T = \text{check}, \ X = e, \ \text{commit}(T), \ X \ \textbf{catch } \_ : \_ \rightarrow \text{rollback}(T), \ e' \ \textbf{end}$$

In this case, we first introduce a checkpoint which reduces to a fresh (unique) identifier, say τ, and saves the current state of the process as a side-effect; variable T is bound to τ. Then, if the evaluation of expression e completes successfully, we gather the computed value in variable X, which is returned after commit(τ) removes the checkpoint.[1] Otherwise, if an exception is raised, the execution jumps to the catch statement and rollback(τ) rolls back the process to the state saved by checkpoint τ (possibly also rolling back other processes in order to get a *causally consistent* state; see below).

Our approach to rollback recovery is based on the notion of *reversible computation* (see [1,13] and references therein). Most programming languages are irreversible, in the sense that the execution of a statement cannot generally be undone. This is the case of Erlang, for instance. Nevertheless, in these languages, one can still define a so-called *Landauer embedding* [19] so that computations become reversible. Intuitively speaking, this operation amounts to defining an instrumented semantics where states carry over a *history* with past states. While this approach may seem impractical at first, there are several useful reversibilization techniques that are roughly based on this idea (typically including some optimization to reduce the amount of saved information, as in, e.g., [28,32]).

While the notion of reversible computation is quite natural in a sequential programming language, the extension to concurrent and distributed languages presents some additional challenges. Danos and Krivine [2] first introduced the notion of *causal consistency* in the context of a reversible calculus (Milner's CCS [30]). Essentially, in a causal consistent reversible setting, *an action cannot be undone until all the actions that causally depend on this action have been already undone.* E.g., we cannot undo the spawning of a process until all the actions of this process have been already undone; similarly, we cannot undo the

---

[1] Binding a temporal variable X to the evaluation of expression e is required so that the try_catch expression still reduces to the same value of the original try_catch expression; if we had just "$T = \text{check}, \ e, \ \text{commit}(T)$" then this sequence would reduce to the value returned by commit in Erlang, thus changing the original semantics.

sending of a message until its reception (and the subsequent actions) have been undone. This notion of causality is closely related with Lamport's "happened before" relation [18]. In our work, we use a similar notion of causality to either propagate checkpoints and to perform causal consistent rollbacks.

Our main contributions are the following. First, we propose the use of three explicit operators for rollback recovery and provide their semantics. They can be used for defining a sort of *safe* transactions, as mentioned above, but not only. For instance, they could also be used as the basis of a reversible debugging scheme where only some computations of interest are reversible, thus reducing the run time overhead. Then, we define a rollback semantics for the extended language that may proceed both as the standard semantics (when no checkpoint is active) or as a reversible semantics (otherwise).

2   A Message-Passing Concurrent Language

In this work, we consider a simple concurrent language that mainly follows the *actor model* [16]. Here, a running system consists of a number of processes (or actors) that can (dynamically) create new processes and can only interact through message sending and receiving (i.e., no shared memory). This is the case, e.g., of (a significant subset of) the functional and concurrent language Erlang [5].

In the following, we will ignore the sequential component of the language (e.g., a typical eager functional programming language in the case of Erlang) and will focus on its concurrent actions:

- *Process spawning.* A process may spawn new processes dynamically. Each process is identified by a *pid* (a shorthand for *p*rocess *id*entifier), which is unique in a running system.
- *Message sending.* A process can send a message to any other process as long as it knows the pid of the target process. This action is asynchronous.
- *Message reception.* Messages need not be immediately consumed by the target process; rather, they are stored in an associated mailbox until they are consumed (if any). We consider so-called *selective* receives, i.e., a process does not necessarily consume the messages in its mailbox in the same order they were delivered, since receive statements may impose additional constraints. When no message matches any constraint, the execution of the process is *blocked* until a matching message reaches its mailbox.

In the following, we let $s, s', \ldots$ denote *states*, typically including some environment, an expression (or statement) to be evaluated and, in some case, a stack. The structure of states is not relevant for the purpose of this paper, though.

A *process* configuration is denoted by a tuple of the form $\langle p, s \rangle$, where p is the pid of the process and s is its current state. Messages have the form (p, p', v) where p is the pid of the sender, p' that of the receiver, and v is the message value. A *system* is then denoted by a parallel composition of both processes and (*floating*) messages, as in [20,26] (instead of using a *global mailbox*, as in [23,25]).

$$(Seq) \quad \frac{s \xrightarrow{\text{seq}} s'}{\langle p, s \rangle \rightarrowtail_{p,\text{seq}} \langle p, s' \rangle}$$

$$(Send) \quad \frac{s \xrightarrow{\text{send}(p',v)} s'}{\langle p, s \rangle \rightarrowtail_{p,\text{send}} (p, p', v) \mid \langle p, s' \rangle}$$

$$(Receive) \quad \frac{s \xrightarrow{\text{rec}(\kappa,cs)} s' \text{ and } \mathsf{matchrec}(cs, v) = cs_i}{(p', p, v) \mid \langle p, s \rangle \rightarrowtail_{p,\text{rec}} \langle p, s'[\kappa \leftarrow cs_i] \rangle \rangle}$$

$$(Spawn) \quad \frac{s \xrightarrow{\text{spawn}(\kappa,s_0)} s' \text{ and } p' \text{ is a fresh pid}}{\langle p, s \rangle \rightarrowtail_{p,\text{spawn}(p')} \langle p, s'[\kappa \leftarrow p'] \rangle \mid \langle p', s_0 \rangle}$$

$$(Par) \quad \frac{S_1 \rightarrowtail_e S_1' \text{ and } id(S_1') \cap id(S_2) = \emptyset}{S_1 \mid S_2 \rightarrowtail_e S_1' \mid S_2}$$

Fig. 1. Standard semantics

A floating message thus represents a message that has been already sent but not yet delivered (i.e., the message is in the network). Furthermore, in this work, process mailboxes are abstracted away for simplicity, thus a floating message can also represent a message that is actually stored in a process mailbox.[2]

Systems range over by S, S', S_1, etc. Here, the parallel composition operator is denoted by "$\mid$" and considered commutative and associative. Therefore, two systems are considered equal if they are the same up to associativity and commutativity.

As in [23,25], the semantics of the language is defined in a modular way, so that the labeled transition relations $\rightarrow$ and $\rightarrowtail$ model the evaluation of expressions (or statements) and the evaluation of systems, respectively.

In the following, we skip the definition of the local semantics ($\rightarrow$) since it is not necessary for our developments; we refer the interested reader to [14,23]. The rules of the operational semantics that define the reduction of systems is shown in Fig. 1. The transition steps are labeled with the pid of the selected process and the considered action: seq, send, rec, or spawn(p'), where p' is the pid of the spawned process. Let us briefly explain these rules:

- Sequential, local steps are dealt with rule *Seq*. Here, we just propagate the reduction from the local level to the system level.
- Rule *Send* applies when the local evaluation requires sending a message as a side effect. The local step $s \xrightarrow{\text{send}(p',v)} s'$ is labeled with the information that must flow from the local level to the system level: the pid of the target process, p', and the message value, v. The system rule then adds a new message of

---

[2] In Erlang, the order of messages sent directly from process p to process p' is preserved when they are all delivered; see [7, Sect. 10.8]. We ignore this constraint for simplicity, but could be ensured by introducing triples of the form (p, p', vs) where vs is a queue of messages instead of a single message.

the form (p, p', v) to the system, where p is the pid of the sender, p' the pid of the target process, and v the message value.

- In order to receive a message, the situation is somehow different. Here, we need some information to flow both from the local level to the system level (the clauses cs of the receive statement) and vice versa (the selected clause cs_i). For this purpose, in rule *Receive*, the label of the local step includes a special variable κ —a sort of *future*— that denotes the position of the receive expression within state s. The rule then checks if there is a floating message v addressed to process p that matches one of the constraints in cs. This is done by the auxiliary function matchrec, which returns the selected clause cs_i of the receive statement in case of a match (the details are not relevant here). Then, the reduction proceeds by binding κ in s' with the selected clause cs_i, which we denote by $s'[\kappa \leftarrow cs_i]$.

- Rule *Spawn* also requires a bidirectional flow of information. Here, the label of the local step includes the future κ and the state of the new process s_0. The rule then produces a fresh pid, p', adds the new process $\langle p', s_0 \rangle$ to the system, and updates the state s' by binding κ to p' (since spawn reduces to the pid of the new process), which we denote by $s'[\kappa \leftarrow p']$.

- Finally, rule *Par* is used to lift an evaluation step to a larger system [26]. The auxiliary function *id* takes a system S and returns the set of pids in S, in order to ensure that new pids are indeed fresh in the complete system.

In the following, $\rightarrowtail^*$ denotes the transitive and reflexive closure of $\rightarrowtail$. Given systems S_0, S_n, we let $S_0 \rightarrowtail^* S_n$ denote a *derivation* under the standard semantics. When we want to consider the individual steps of a derivation, we often write $S_0 \rightarrowtail_{p_1,a_1} S_1 \rightarrowtail_{p_2,a_2} \dots \rightarrowtail_{p_n,a_n} S_n$. A reduction step usually consists of a number of applications of rule *Par* until a process, or a combination of a process and a message, is selected, so that one of the remaining rules can be applied (*Seq*, *Send*, *Receive* or *Spawn*). We often omit the steps with rule *Par* and only show the reductions on the selected process, i.e., $a_i \in \{\text{seq}, \text{send}, \text{rec}, \text{spawn}(p')\}$.

An *initial* system has the form $\langle p, s_0 \rangle$, i.e., it contains a single process. A system S' is *reachable* if there exists a derivation $S \rightarrowtail^* S'$ such that S is an initial system. A derivation $S \rightarrowtail^* S'$ is *well-defined* under the standard semantics if S is a reachable system.

The semantics in Fig. 1 applies to a significant subset of the programming language Erlang [5], as described, e.g., in [14,25]. However, for clarity, we will consider in the examples a much simpler notation which only shows some relevant information regarding the concurrent actions performed by each process. This is just a textual representation which makes explicit process interaction but is not the actual program. In particular, we describe the concurrent actions of a process by means of the following items:

- $p \leftarrow \text{spawn}()$, for process spawning, where p is the (fresh) pid returned by the call to spawn and assigned to the new process;

- send(p, v), for sending a message, where p is the pid of the target process and v the message value (which could be a tuple including the process own pid in order to get a reply, a common practice in Erlang);
- rec(v), for receiving message v.

We will ignore sequential actions in the examples since they are not relevant for the purpose of this paper.

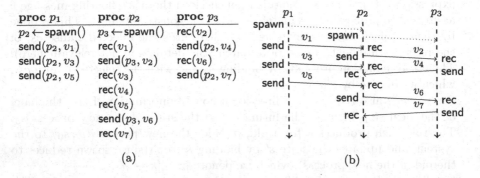

proc p_1	**proc** p_2	**proc** p_3
$p_2 \leftarrow$ spawn()	$p_3 \leftarrow$ spawn()	rec(v_2)
send(p_2, v_1)	rec(v_1)	send(p_2, v_4)
send(p_2, v_3)	send(p_3, v_2)	rec(v_6)
send(p_2, v_5)	rec(v_3)	send(p_2, v_7)
	rec(v_4)	
	rec(v_5)	
	send(p_3, v_6)	
	rec(v_7)	

(a)

(b)

Fig. 2. Textual and graphical representation of the concurrent actions of a program (time flows from top to bottom)

Example 1. Consider, for instance, a system with three processes with pids p_1 (the initial one), p_2, and p_3, which perform the actions shown in Fig. 2 (a). A (partial) derivation under the standard semantics (representing a particular interleaving of the processes' actions) may proceed as follows, where we underline the selected process and message (if any) at each step:

$$\langle \underline{p_1}, s[p_2 \leftarrow \mathsf{spawn}()] \rangle$$
$$\longmapsto_{p_1,\mathsf{spawn}(p_2)} \langle \underline{p_1}, s[\mathsf{send}(p_2, v_1)] \rangle \mid \langle p_2, s[p_3 \leftarrow \mathsf{spawn}()] \rangle$$
$$\longmapsto_{p_2,\mathsf{spawn}(p_3)} \langle \underline{p_1}, s[\mathsf{send}(p_2, v_1)] \rangle \mid \langle p_2, s[\mathsf{rev}(v_1)] \rangle \mid \langle p_3, s[\mathsf{rec}(v_2)] \rangle$$
$$\longmapsto_{p_1,\mathsf{send}} \langle p_1, s[\mathsf{send}(p_2, v_3)] \rangle \mid \langle \underline{p_2}, s[\mathsf{rec}(v_1)] \rangle \mid \langle p_3, s[\mathsf{rec}(v_2)] \rangle \mid \underline{(p_1, p_2, v_1)}$$
$$\longmapsto_{p_2,\mathsf{rec}} \langle p_1, s[\mathsf{send}(p_2, v_3)] \rangle \mid \langle \underline{p_2}, s[\mathsf{send}(p_3, v_2)] \rangle \mid \langle p_3, s[\mathsf{rec}(v_2)] \rangle$$
$$\longmapsto_{p_2,\mathsf{send}} \langle p_1, s[\mathsf{send}(p_2, v_3)] \rangle \mid \langle \underline{p_2}, s[\mathsf{rec}(v_3)] \rangle \mid \langle p_3, s[\mathsf{rec}(v_2)] \rangle \mid \underline{(p_2, p_3, v_2)}$$
$$\longmapsto_{p_3,\mathsf{rec}} \langle \underline{p_1}, s[\mathsf{send}(p_2, v_3)] \rangle \mid \langle p_2, s[\mathsf{rec}(v_3)] \rangle \mid \langle p_3, s[\mathsf{send}(p_2, v_4)] \rangle$$
$$\longmapsto_{p_1,\mathsf{send}} \langle p_1, s[\mathsf{send}(p_2, v_5)] \rangle \mid \langle p_2, s[\mathsf{rec}(v_3)] \rangle \mid \langle p_3, s[\mathsf{send}(p_2, v_4)] \rangle \mid \underline{(p_1, p_2, v_3)}$$
$$\longmapsto \ldots$$

where a state of the form $s[op]$ denotes an arbitrary state where the next operation to be reduced is op. We omit some intermediate steps which are not relevant here. A graphical representation of the reduction can be found in Fig. 2 (b).

We note that programs can exhibit an iterative behavior through recursive function calls. E.g., a typical server process is defined by a function that waits for a client request, process it, and then makes a recursive call, possibly with a modified *state* (a common pattern in Erlang). This is hidden in our examples since we do not show sequential operations explicitly.

$$(Check) \ \frac{}{\theta, \mathsf{check} \ \xrightarrow{\mathsf{check}(\kappa)} \ \theta, \kappa}$$

$$(Commit) \ \frac{}{\theta, \mathsf{commit}(\tau) \ \xrightarrow{\mathsf{commit}(\tau)} \ \theta, ok} \qquad (Rollback) \ \frac{}{\theta, \mathsf{rollback}(\tau) \ \xrightarrow{\mathsf{rollback}(\tau)} \ \theta, ok}$$

Fig. 3. Rollback recovery operators

3   Checkpoint-Based Rollback Recovery

In this section, we present our approach to rollback recovery in a message-passing concurrent language. Essentially, our approach is based on defining an instrumented semantics with two modes: a "normal" mode, which proceeds similarly to the standard semantics, and a "reversible" mode, where actions can be undone and, thus, can be used for rollbacks.

3.1   Basic Operators

We consider three explicit operators to control rollback recovery: check, commit, and rollback. Intuitively speaking, they proceed as follows:

- check introduces a *checkpoint* for the current process. The reduction of check returns a fresh identifier, τ, associated to the checkpoint; note that nested checkpoints are possible.
- commit(τ) can then be used to discard the state saved in checkpoint τ. In our context, check implies turning the reversible mode on and commit turning it off (when no more active checkpoints exist).
- Finally, rollback(τ) starts a backward computation, undoing all the actions of the process (and their causal dependencies from other processes) up to the call to check (including) that introduced τ.

The local reduction rules for the new operators are very simple and can be found in Fig. 3. Here, we consider that a local state consists of an environment (a variable substitution) and an expression (to be evaluated), but it could be straightforwardly extended to other state configurations (e.g., configurations that also include a stack, as in [15]).

Rule *Check* reduces the call to a future, κ, which also occurs in the label of the transition step. As we will see in the next section, the corresponding rule in the system semantics will perform the associated side-effect (creating a checkpoint) and will also bind κ with the (fresh) identifier for this checkpoint. Rules *Commit* and *Rollback* just pass the corresponding information to the system semantics in order to do the associated side effects. Both rules reduce the call to the constant "*ok*" (an atom used in Erlang when a function call does not return any value).

Example 2. Consider again a program that performs the concurrent actions of Example 1, where we now add a couple of checkpoints, a commit, and a rollback to process p_1, as shown in Fig. 4. Here, we let $\tau \leftarrow$ check denote that τ is the

proc p_1	proc p_2	proc p_3
$p_2 \leftarrow \mathsf{spawn}()$	$p_3 \leftarrow \mathsf{spawn}()$	$\mathsf{rec}(v_2)$
$\mathsf{send}(p_2, v_1)$	$\mathsf{rec}(v_1)$	$\mathsf{send}(p_2, v_4)$
$\tau_1 \leftarrow \mathsf{check}$	$\mathsf{send}(p_3, v_2)$	$\mathsf{rec}(v_6)$
$\mathsf{send}(p_2, v_3)$	$\mathsf{rec}(v_3)$	$\mathsf{send}(p_2, v_7)$
$\tau_2 \leftarrow \mathsf{check}$	$\mathsf{rec}(v_4)$	
$\mathsf{send}(p_2, v_5)$	$\mathsf{rec}(v_5)$	
$\mathsf{commit}(\tau_2)$	$\mathsf{send}(p_3, v_6)$	
$\mathsf{rollback}(\tau_1)$	$\mathsf{rec}(v_7)$	

Fig. 4. Concurrent actions of a program including check, commit, and rollback

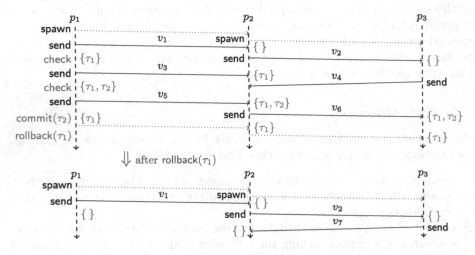

Fig. 5. Graphical representation: before and after the rollback

(fresh) identifier returned by the call to check. A graphical representation of a particular execution can be found in Fig. 5. Intuitively speaking, it proceeds as follows:

- Process p_1 calls function check, which creates a checkpoint with identifier τ_1. This checkpoint is propagated to p_2 when sending message v_3, so p_2 turns the reversible mode on too.
- Then, p_1 creates another checkpoint, τ_2, so we have two active checkpoints. These checkpoints are propagated to p_2 by message v_5 and also to p_3 by message v_6. At this point, all three processes have the reversible mode on.
- Now, p_1 calls $\mathsf{commit}(\tau_2)$, so checkpoint τ_2 is not active anymore in p_1. This is also propagated to both p_1 and p_2. Nevertheless, the reversible mode is still on in all three processes since τ_1 is still alive.
- Then, p_1 calls $\mathsf{rollback}(\tau_1)$ so p_1 undoes all its actions up to (and including) the first call to check. For this rollback to be causal consistent, p_2 rolls back to the point immediately before receiving message v_3, and p_3 to the point immediately before receiving message v_6.

- Finally, all three processes are back to normal, irreversible mode (no checkpoint is active), and p_3 sends message v_7 to p_2. The actions in the second diagram are all irreversible.

3.2 A Reversible Semantics for Rollback Recovery

Let us now consider the design of a *reversible* semantics for rollback recovery. Essentially, the operators can be modeled as follows:

- The reduction of check creates a checkpoint, which turns on the reversible mode of a process as a side-effect (assuming it was not already on). As in [23,25], reversibility is achieved by defining an appropriate *Landauer embedding* [19], i.e., by adding a history of the computation to each process configuration.[3] A checkpoint is propagated to other processes when a causally dependent action is performed (i.e., spawn and send); following the terminology of [3], these checkpoints are called *forced* checkpoints.
- A call of the form commit(τ) removes τ from the list of *active* checkpoints of a process, turning the reversible mode off when the list of active checkpoints is empty. Forced checkpoints in other processes with the same identifier τ (if any) are also removed from the corresponding sets of active checkpoints.
- Finally, the reduction of rollback(τ) involves undoing all the steps of a given process up to the checkpoint τ *in a causal consistent way*, i.e., possibly also undoing causally dependent actions from other processes.

Unfortunately, this apparently simple model presents a problem. Consider, for instance, a process p that performs the following actions:

$$p := \{\overbrace{\mathsf{check}(\tau_1), \mathsf{send}(p', v), \underbrace{\mathsf{check}(\tau_2), \dots, \mathsf{commit}(\tau_1)}, \dots, \mathsf{rollback}(\tau_2)}, \dots\}$$

$$(*)$$

Here, we can observe that the pairs check-commit and check-rollback are not well balanced. As a consequence, we commit checkpoint τ_1 despite the fact that a rollback like rollback(τ_2) may bring the computation back to the point immediately before check(τ_2), where τ_1 should be alive, thus producing an inconsistent state. We could recover τ_1 when undoing the call commit(τ_1). However, this is not only a *local* problem, since commit(τ_1) may have also removed the checkpoint from other processes (p', in the example).

We could solve the above problem by considering that a call to commit may introduce new causal dependencies. Intuitively speaking, one could treat the propagation of commit to other process as the sending of a message. Therefore, a causal consistent rollback would often require undoing all subsequent actions in these processes before undoing the call to commit. Unfortunately, this solution

---

[3] For clarity of exposition, the complete state is saved at each state. Nevertheless, an optimized history could also be defined; see, e.g., [28,32].

would require all processes to keep the reversible mode on all the time, which is precisely what we want to avoid.

A very simple workaround comes immediately to mind: require the programmer to write well-balanced pairs check-commit and check-rollback. In this case, a rollback that undo a call to commit will also undo the corresponding call to check, so the inconsistent situation above would no longer be possible. However, this solution is not acceptable when forced checkpoints come into play. For instance, in example (*) above, if we replace check(τ_2) by a receive operation from some process where checkpoint τ_2 is active, the same inconsistent state can be reproduced. In this case, though, we cannot require the programmer to avoid such situations since they are unpredictable.

For all the above, in this work we do not impose any constraint on the use of the new operators but propose the following solution: When a call of the form commit(τ) occurs, we check in the process' history whether τ is the *last* active checkpoint of the process (either proper or forced). If this is the case, the commit is executed. Otherwise, it is *delayed* until the condition is met.

Our configurations will now have three additional fields: the set of active checkpoints, the set of delayed commits, and a history (for reversibility):

Definition 1 (rollback configuration). *A forward configuration is defined as a tuple $(\mathcal{C}, \mathcal{D}, h, \langle p, s \rangle)$ where $\mathcal{C}$ is a set of (active) checkpoint identifiers, $\mathcal{D}$ is a set of delayed commits, h is a history, and $\langle p, s \rangle$ is a pid and a (local) state, similarly to the standard semantics. A backward configuration has the form $((\mathcal{C}, \mathcal{D}, h, \langle p, s \rangle))^\tau$ where τ is the target of a rollback request.*

A (rollback) configuration is either a forward or a backward configuration.

As for the *messages*, they now have the form $(\mathcal{C}, p, p', \{\ell, v\})$. Here, we can distinguish two differences w.r.t. the standard semantics: first, we add a set of active checkpoints, $\mathcal{C}$, which should be propagated to the receiver as *forced* checkpoints; secondly, the message value is wrapped with a tag to uniquely identify it (as in [23], in order to distinguish messages with the same value).

As in the standard semantics, a *system* is then a parallel composition of (rollback) configurations and floating messages.

In the following, we let [ ] denote an empty list and $x : xs$ a list with head x and tail xs. A process *history* h is then represented by a list of the following elements: seq, send, rec, spawn, check, and commit. Each one is denoted by a term containing enough information to undo the corresponding reduction step,[4] except for the case of commit, whose side-effects are irreversible, as argued above. To be more precise,

- all terms store the current state (s).
- for sending a message, the corresponding term also stores the pid of the target process (p') and the message tag (ℓ);
- for receiving a message, the term also stores two sets of active checkpoints (the active ones and those received from a message as forced checkpoints), the pid of the sender (p'), the message tag (ℓ), and the message value (v);

---

[4] The reader can compare the rules in Fig. 6 and their inverse counterpart in Fig. 7.

- for spawning a process, the corresponding term also includes the fresh pid of the new process (p');
- and, finally, for check and commit, it also stores the checkpoint identifier (τ).

A delayed commit is represented as a triple $\langle \tau, h, P \rangle$, where τ is a checkpoint identifier, h is a history, and P is a set of pids (the pids of the processes where a forced checkpoint τ has been propagated).

$$(Seq) \quad \frac{s \xrightarrow{\text{seq}} s' \text{ and } add_{\mathcal{C}}(\text{seq}(s), h) = h'}{(\mathcal{C}, \mathcal{D}, h, \langle p, s \rangle) \hookrightarrow_{p,\text{seq}} (\mathcal{C}, \mathcal{D}, h', \langle p, s' \rangle)}$$

$$(Send) \quad \frac{s \xrightarrow{\text{send}(p',v)} s', \ \ell \text{ is a fresh symbol, and } add_{\mathcal{C}}(\text{send}(s, p', \ell), h) = h'}{(\mathcal{C}, \mathcal{D}, h, \langle p, s \rangle) \hookrightarrow_{p,\text{send}(\ell)} (\mathcal{C}, p, p', \{\ell, v\}) \mid (\mathcal{C}, \mathcal{D}, h', \langle p, s' \rangle)}$$

$$(Receive) \quad \frac{s \xrightarrow{\text{rec}(\kappa,cs)} s', \ \text{matchrec}(cs, v) = cs_i, \text{ and } add_{\mathcal{C}}(\text{rec}(\mathcal{C}' \setminus \mathcal{C}, \mathcal{C}', s, p', \ell, v), h) = h'}{(\mathcal{C}', p', p, \{\ell, v\}) \mid (\mathcal{C}, \mathcal{D}, h, \langle p, s \rangle) \hookrightarrow_{p,\text{rec}(\ell)} (\mathcal{C} \cup \mathcal{C}', \mathcal{D}, h', \langle p, s'[\kappa \leftarrow cs_i] \rangle)}$$

$$(Spawn) \quad \frac{s \xrightarrow{\text{spawn}(\kappa,s_0)} s', \ p' \text{ is a fresh pid, and } add_{\mathcal{C}}(\text{spawn}(s, p'), h) = h'}{(\mathcal{C}, \mathcal{D}, h, \langle p, s \rangle) \hookrightarrow_{p,\text{spawn}(p')} (\mathcal{C}, \mathcal{D}, h', \langle p, s'[\kappa \leftarrow p'] \rangle) \mid (\mathcal{C}, [\,], \langle p', s_0 \rangle)}$$

$$(Check) \quad \frac{s \xrightarrow{\text{check}(\kappa)} s' \text{ and } \tau \text{ is a fresh identifier}}{(\mathcal{C}, \mathcal{D}, h, \langle p, s \rangle) \hookrightarrow_{p,\text{check}(\tau)} (\mathcal{C} \cup \{\tau\}, \mathcal{D}, \text{check}(\tau, s):h, \langle p, s'[\kappa \leftarrow \tau] \rangle)}$$

$$(Commit) \quad \frac{s \xrightarrow{\text{commit}(\tau)} s', \ last_\tau(h) = true, \ dp_\tau(h) = P, \text{ and } propagate(\tau, P)}{(\mathcal{C}, \mathcal{D}, h, \langle p, s \rangle) \hookrightarrow_{p,\text{commit}(\tau)} (\mathcal{C} \setminus \{\tau\}, \mathcal{D}, \text{commit}(\tau, s):h, \langle p, s' \rangle)}$$

$$\frac{s \xrightarrow{\text{commit}(\tau)} s', \ last_\tau(h) = false, \text{ and } dp_\tau(h) = P}{(\mathcal{C}, \mathcal{D}, h, \langle p, s \rangle) \hookrightarrow_{p,\text{delay}(\tau)} (\mathcal{C} \setminus \{\tau\}, \mathcal{D} \cup \{\langle \tau, h, P \rangle\}, \text{commit}(\tau, s):h, \langle p, s' \rangle)}$$

$$(Delay) \quad \frac{last_\tau(h') = true \text{ and } propagate(\tau, P)}{(\mathcal{C}, \mathcal{D} \cup \{\langle \tau, h', P \rangle\}, h, \langle p, s \rangle) \hookrightarrow (\mathcal{C}, \mathcal{D}, h, \langle p, s \rangle)}$$

$$(Rollback) \quad \frac{s \xrightarrow{\text{rollback}(\tau)} s'}{(\mathcal{C}, \mathcal{D}, h, \langle p, s \rangle) \hookrightarrow_{p,\text{rollback}(\tau)} ((\mathcal{C}, \mathcal{D}, h, \langle p, s' \rangle))^\tau}$$

$$(Par) \quad \frac{S_1 \hookrightarrow_l S_1' \text{ and } id(S_1') \cap id(S_2) = \emptyset}{S_1 \mid S_2 \hookrightarrow_l S_1' \mid S_2}$$

Fig. 6. Rollback recovery semantics: forward rules

Forward Rules. The *forward* reduction rules of the rollback semantics are shown in Fig. 6. The main difference with the standard semantics is that, now, some process configurations include a *history* with enough information to undo any reduction step. Here, we follow the same strategy as in [23, 25] in order to define a scheme for reversible debugging.[5]

---

[5] In contrast to [23, 25], however, we do not need to undo every possible step, but only those steps that are performed when there is at least one active checkpoint. This is why we added $\mathcal{C}$: to store the set of *active* checkpoints (i.e., checkpoints without a corresponding commit/rollback yet). Observe that we might have several active checkpoints not only because nested checkpoints are possible, but because of *forced* checkpoints propagated by process spawning and message sending.

Essentially, the first four rules of the semantics can behave either as the standard semantics or as a *reversible* semantics, depending on whether $\mathcal{C}$ is empty or not. For conciseness, we avoid duplicating all rules by introducing the auxiliary function add to update the history only when there are active checkpoints: $\mathsf{add}_{\mathcal{C}}(a, h) = h$ if $\mathcal{C} = \emptyset$, and $\mathsf{add}_{\mathcal{C}}(a, h) = a\!:\!h$ otherwise.

As mentioned above, the reversible mode is propagated through message sending and receiving. This is why messages now include the set of active checkpoints $\mathcal{C}$. As can be seen in rule *Receive*, the process receiving the message updates its active checkpoints with those in the message. This is necessary for rollbacks to be causally consistent. Note that, in the associated term in the history, $\mathsf{rec}(\mathcal{C}' \setminus \mathcal{C}, \mathcal{C}', s, p', \ell, v)$, $\mathcal{C}' \setminus \mathcal{C}$ denotes the *forced* checkpoints introduced by the received message.

Similarly, the reversible mode is also propagated by process spawning: rule *Spawn* adds the current set of active checkpoints $\mathcal{C}$ (which might be empty) to the new process.

As for the new rules, *Check* produces a fresh identifier, τ, and binds the future, κ, to this identifier. It also adds τ to the current set of active checkpoints. Note that, if $\mathcal{C}$ is empty, this step turns the reversible mode on.

Commit includes two transition rules, depending on whether the commit can be done or it should be delayed. We use the auxiliary Boolean function *last* so that $last_{\tau}(h)$ checks whether τ is the last checkpoint of the process according to history h, i.e., whether the last check or rec term in h has either the form $\mathsf{check}(\tau, s')$ or $\mathsf{rec}(\mathcal{C}', \ldots)$ with $\tau \in \mathcal{C}'$. Note that we do not need to consider forced checkpoints introduced by process spawning since they cannot occur after a call to check (they are always introduced when spawning the process). If the call to function *last* returns *true*, we remove the checkpoint identifier from $\mathcal{C}$ and from all processes where this checkpoint was propagated (as a forced checkpoint). Here, the auxiliary function dp_{τ} takes a history and returns all pids which have causal dependencies with the current process according to h, i.e.,

- $p \in dp_{\tau}(h)$ if $\mathsf{spawn}(s, p)$ occurs in h;
- $p \in dp_{\tau}(h)$ if $\mathsf{send}(s, p, \ell)$ occurs in h.

Now, we want to propagate the effect of commit to all processes in $dp_{\tau}(h)$ in order to remove τ from their set of active checkpoints. One could formalize this process with a few more transition rules and a new kind of configuration. However, for simplicity, we represent it by means of an auxiliary function *propagate* so that $propagate(\tau, P)$ always returns *true* and performs the following side-effects:

1. for each $p' \in P$, we look for the process with pid p', say $(\mathcal{C}', \mathcal{D}', h', \langle p', s' \rangle)$;
2. if $\tau \notin \mathcal{C}'$ (it is not an active checkpoint of process p'), we are done;
3. otherwise ($\tau \in \mathcal{C}'$), we remove τ from $\mathcal{C}'$ and repeat the process, i.e., we compute $P' = dp_{\tau}(h')$ and call $propagate(\tau, P')$.

Termination is ensured since the number of processes is finite and a process where τ is not active will eventually be reached. In practice, commit propagation can be implemented by sending (asynchronous) messages to the involved processes.

$(\overline{Seq})$ $\qquad$ $((C, \mathcal{D}, \mathsf{seq}(s) : h, \langle p, s' \rangle))^\tau \hookrightarrow_{p, \overline{\mathsf{seq}}} ((C, \mathcal{D}, h, \langle p, s \rangle))^\tau$

$(\overline{Send})$ $\quad (C, p, p', \{\ell, v\}) \mid ((C, \mathcal{D}, \mathsf{send}(s, p', \ell) : h, \langle p, s' \rangle))^\tau \hookrightarrow_{p, \overline{\mathsf{send}}(\ell)} ((C, \mathcal{D}, h, \langle p, s \rangle))^\tau$

$\qquad ((C, \mathcal{D}, \mathsf{send}(s, p', \ell) : h, \langle p, s' \rangle))^\tau \mid (C', \mathcal{D}', h', \langle p', s'' \rangle)$
$\qquad \hookrightarrow ((C, \mathcal{D}, \mathsf{send}(s, p', \ell) : h, \langle p, s' \rangle))^\tau \mid ((C', \mathcal{D}', h', \langle p', s'' \rangle))^\tau$

$(\overline{Receive})$ $\quad ((C \cup C', \mathsf{rec}(C'', C', s, p', \ell, v) : h, \langle p, s' \rangle))^\tau$
$\qquad\qquad \hookrightarrow_{p, \overline{\mathsf{rec}}(\ell)} (C', p', p, \{\ell, v\}) \mid (C, \mathcal{D}, h, \langle p, s \rangle) \quad \text{if } \tau \in C''$

$\qquad ((C \cup C', \mathsf{rec}(C'', C', s, p', \ell, v) : h, \langle p, s' \rangle))^\tau$
$\qquad\qquad \hookrightarrow_{p, \overline{\mathsf{rec}}(\ell)} (C', p', p, \{\ell, v\}) \mid ((C, \mathcal{D}, h, \langle p, s \rangle))^\tau \quad \text{if } \tau \notin C''$

$(\overline{Spawn})$ $\quad ((C, \mathcal{D}, \mathsf{spawn}(s, p') : h, \langle p, s' \rangle))^\tau \mid ((C, \emptyset, [\,], \langle p', s_0 \rangle))^\tau \hookrightarrow_{p, \overline{\mathsf{spawn}}(p')} ((C, \mathcal{D}, h, \langle p, s \rangle))^\tau$

$\qquad ((C, \mathcal{D}, \mathsf{spawn}(s, p') : h, \langle p, s' \rangle))^\tau \mid (C', \mathcal{D}', h', \langle p', s'' \rangle)$
$\qquad \hookrightarrow ((C, \mathcal{D}, \mathsf{spawn}(s, p') : h, \langle p, s' \rangle))^\tau \mid ((C', \mathcal{D}', h', \langle p', s'' \rangle))^\tau$

$(\overline{Check})$ $\quad ((C, \mathcal{D}, \mathsf{check}(\tau, s) : h, \langle p, s' \rangle))^\tau \hookrightarrow_{p, \overline{\mathsf{check}}(\tau)} (C \setminus \{\tau\}, \mathcal{D}, h, \langle p, s \rangle)$

$\qquad ((C, \mathcal{D}, \mathsf{check}(\tau', s) : h, \langle p, s' \rangle))^\tau \hookrightarrow_{p, \overline{\mathsf{check}}(\tau')} ((C \setminus \{\tau'\}, \mathcal{D}, h, \langle p, s \rangle))^\tau \quad \text{if } \tau \neq \tau'$

$(\overline{Commit})$ $\quad ((C, \mathcal{D}, \mathsf{commit}(\tau', s) : h, \langle p, s' \rangle))^\tau$
$\qquad\qquad \hookrightarrow_{p, \overline{\mathsf{commit}}(\tau')} ((C \cup \{\tau'\}, \mathcal{D}, h, \langle p, s \rangle))^\tau \quad \text{if } \langle \tau', \_, \_ \rangle \notin \mathcal{D}$

$\qquad ((C, \mathcal{D} \cup \{\langle \tau', h', P \rangle\}, \mathsf{commit}(\tau', s) : h, \langle p, s' \rangle))^\tau$
$\qquad\qquad \hookrightarrow_{p, \overline{\mathsf{commit}}(\tau')} ((C \cup \{\tau'\}, \mathcal{D}, h, \langle p, s \rangle))^\tau$

(*) We assume the side condition $\tau \in C$ holds in all rules.
(**) The second rule of $\overline{Send}$ only applies when the message tagged with ℓ has been received by p' according to history h'.

Fig. 7. Rollback recovery semantics: backward rules

Note that the semantics would be sound even if commit operations were not propagated, so doing it is essentially a matter of efficiency (a sort of *garbage collection* to avoid recording actions that are not really necessary).

On the other hand, if the call to function *last* returns *false*, the checkpoint is moved from C to $\mathcal{D}$ as a delayed commit (second rule of *Commit*). Eventually, rule *Delay* becomes applicable and proceeds similarly to the first rule of *Commit* but considering the delayed commit. We do not formalize a particular strategy for firing rule *Delay*, but a simple strategy would only fire this rule only when some checkpoint is removed from the set of active checkpoints of a process.

Rule *Rollback* simply changes the forward configuration to a backward configuration, also adding the superscript τ to *drive* the rollback. Therefore, the forward rules are no longer applicable to this process (and the backward rules in Fig. 7 can be applied instead).

Finally, rule *Par* is identical to that in the standard semantics. The only difference is that, now, function $id(S)$ returns the set of pids, message tags, and checkpoints in S.

Backward Rules. Let us now present the backward rules of the rollback semantics, which are shown in Fig. 7.

First, rule $\overline{Seq}$ applies when the history is headed by a term of the form $\mathsf{seq}(s)$. It simply removes this element from the history and recovers state s.

Rule $\overline{Send}$ distinguishes two cases. If the message with tag ℓ is a floating message (so it has not been *received*), then we remove the message from the system and recover the saved state. Otherwise (i.e., the message has been consumed by the target process p'), the rollback mode is propagated to process p', which will go backwards up to the receiving of the message; once the floating message is back into the system, the first rule applies.

Rule $\overline{Receive}$ also distinguishes two cases. In both of them, the message is put back into the system as a floating message and the recorded state is recovered. They differ in that the first rule considers the case where τ is a forced checkpoint introduced by the received message. In this case, we undo the step and the process resumes its forward computation.[6] Otherwise (i.e., τ was introduced somewhere else), we undo the step but keep the rollback mode for the process.

Rule $\overline{Spawn}$ proceeds in a similar way as rule $\overline{Send}$: if the spawned process is already in its initial state with an empty history, it is simply removed from the system. Otherwise, the reversible mode is propagated to the spawned process p'.

Rule $\overline{Check}$ applies when we reach a checkpoint in the process' history. If the checkpoint has the same identifier of the initial rollback operator, τ, the job is done and the process resumes its forward computation after undoing one last step.[7] Otherwise (i.e., the checkpoint in the history has a different identifier, τ'), we undo the step, also removing τ' from the set of active checkpoints, but keep the rollback mode.

Finally, rule $\overline{Commit}$ considers two cases: either the commit has been executed (and, thus, the rollback will eventually undo the associated check too), or the commit was delayed (see Example 3 below).

Example 3. Consider again the program in Example 2, where we now switch the arguments of commit and rollback in process p_1 in order to illustrate the use of delayed commits (p_2 and p_3 remain the same as before). The concurrent actions of the modified program are shown in Fig. 8, where the terms send and rec now include message tags instead of values. In this case, the sequence of configurations of p_1 would be as shown in Fig. 9. Here, the call $\mathsf{commit}(\tau_1)$ cannot be executed since the last checkpoint of process p_1 is τ_2. Therefore, it is added as a delayed

---

[6] Here, we assume that rule $\overline{Send}$ has a higher priority than rule $\overline{Receive}$, so once a message is put back into the network, the corresponding message sending is undone (rather than being received again).

[7] We note that, in its current formulation, we would recover the state immediately before the checkpoint and, then, would perform the same actions—up to the non-determinism of the language—. If the goal was to implement a safer try_catch (as illustrated in Sect. 1), then we could slightly modify the rules so that, when the rollback is done, we update the recovered state by replacing the next expression to be evaluated by the expression after the call to rollback (that in s' in rule *Rollback*). We leave this particular extension as future work.

proc p_1	**proc** p_2	**proc** p_3
$p_2 \leftarrow$ spawn()	$p_3 \leftarrow$ spawn()	rec(ℓ_2)
send(p_2, ℓ_1)	rec(ℓ_1)	send(p_2, ℓ_4)
$\tau_1 \leftarrow$ check	send(p_3, ℓ_2)	rec(ℓ_6)
send(p_2, ℓ_3)	rec(ℓ_3)	send(p_2, ℓ_7)
$\tau_2 \leftarrow$ check	rec(ℓ_4)	
send(p_2, ℓ_5)	rec(ℓ_5)	
commit(τ_1)	send(p_3, ℓ_6)	
rollback(τ_2)	rec(ℓ_7)	

Fig. 8. Concurrent actions of the program in Example 3

$(\emptyset, \emptyset, [\,], \langle p_1, s[p_2 \leftarrow \mathsf{spawn}] \rangle)$
$\hookrightarrow (\emptyset, \emptyset, [\mathsf{spawn}(p_2)], \langle p_1, s[\mathsf{send}(p_2, \ell_1)] \rangle)$
$\hookrightarrow (\emptyset, \emptyset, [\mathsf{send}(p_2, \ell_1), \mathsf{spawn}(p_2)], \langle p_1, s[\tau_1 \leftarrow \mathsf{check}] \rangle)$
$\hookrightarrow (\{\tau_1\}, \emptyset, [\mathsf{check}(\tau_1), \mathsf{send}(p_2, \ell_1), \mathsf{spawn}(p_2)], \langle p_1, s[\mathsf{send}(p_2, \ell_3)] \rangle)$
$\hookrightarrow (\{\tau_1\}, \emptyset, [\mathsf{send}(p_2, \ell_3), \mathsf{check}(\tau_1), \mathsf{send}(p_2, \ell_1), \ldots], \langle p_1, s[\tau_2 \leftarrow \mathsf{check}] \rangle)$
$\hookrightarrow (\{\tau_1, \tau_2\}, \emptyset, [\mathsf{check}(\tau_2), \mathsf{send}(p_2, \ell_3), \mathsf{check}(\tau_1), \ldots], \langle p_1, s[\mathsf{send}(p_2, \ell_5)] \rangle)$
$\hookrightarrow (\{\tau_1, \tau_2\}, \emptyset, [\mathsf{send}(p_2, \ell_5), \mathsf{check}(\tau_2), \mathsf{send}(p_2, \ell_3), \ldots], \langle p_1, s[\mathsf{commit}(\tau_1)] \rangle)$
$\hookrightarrow (\{\tau_2\}, \{\langle \tau_1, h, \{p_2\} \rangle\}, [\mathsf{commit}(\tau_1), \mathsf{send}(p_2, \ell_5), \ldots], \langle p_1, s[\mathsf{rollback}(\tau_2)] \rangle)$
$\hookrightarrow ((\{\tau_2\}, \{\langle \tau_1, h, \{p_2\} \rangle\}, [\mathsf{commit}(\tau_1), \mathsf{send}(p_2, \ell_5), \ldots], \langle p_1, s[\ldots] \rangle))^{\tau_2}$
$\hookrightarrow ((\{\tau_1, \tau_2\}, \emptyset, [\mathsf{send}(p_2, \ell_5), \mathsf{check}(\tau_2), \mathsf{send}(p_2, \ell_3), \ldots], \langle p_1, s[\ldots] \rangle))^{\tau_2}$
$\hookrightarrow ((\{\tau_1, \tau_2\}, \emptyset, [\mathsf{check}(\tau_2), \mathsf{send}(p_2, \ell_3), \mathsf{check}(\tau_1), \ldots], \langle p_1, s[\ldots] \rangle))^{\tau_2}$
$\hookrightarrow (\{\tau_1\}, \emptyset, [\mathsf{send}(p_2, \ell_3), \mathsf{check}(\tau_1), \mathsf{send}(p_2, \ell_1), \mathsf{spawn}(p_2)], \langle p_1, s[\ldots] \rangle)$
$\hookrightarrow \ldots$

Fig. 9. Sequence of configurations of p_1 (Example 3), where some arguments of history items and other information (not relevant for the example) are omitted

checkpoint. Then, we have a call rollback(τ_2) which undo the last steps of p_1 (as well as some steps in p_2 and p_3 in order to keep causal consistency, which we do not show for simplicity).

Soundness. In the following, we assume a *fair* selection strategy for processes, so that each process is eventually reduced. Furthermore, we only consider *well-defined* derivations where the calls commit(τ) and rollback(τ) can only be made by the same process that created the checkpoint τ, and a process can only have one action for every checkpoint τ, either commit(τ) or rollback(τ), but not both.

Soundness is then proved by projecting the configurations of the rollback semantics to configurations of either the standard semantics (function *sta*) or a *pure* reversible semantics (function *rev*). Then, we prove that every step under the rollback semantics has a counterpart either under the standard or under the reversible semantics, after applying the corresponding projections. Formally,[8]

---

[8] We denote by $\rightarrow^=$ the reflexive closure of a binary relation $\rightarrow$, i.e., $(\rightarrow^=) = (\rightarrow \cup =)$. We consider the reflexive closure in the claim of Theorem 1 since some steps under the rollback semantics have no counterpart under the standard or reversible semantics. In these cases, the projected configurations remain the same.

Theorem 1. *Let d be a well-defined derivation under the rollback semantics. For each step $S \hookrightarrow S'$ in d we have either $sta(S) \rightarrowtail^= sta(S')$ or $rev(S) \rightleftharpoons^= rev(S')$.*

We have also proved that every computation between a checkpoint and the corresponding rollback is indeed reversible for well-defined derivations; see the companion technical report, [37], for more details. We leave the study of other interesting results of our rollback semantics (e.g., minimality and some partial completeness) for future work.

4   Related Work

There is abundant literature on checkpoint-based rollback recovery to improve fault tolerance (see, e.g., the survey by Elnozahy *et al* [3] and references there in). In contrast to most of these approaches, our distinctive features are the extension of the underlying language with *explicit* operators for rollback recovery, the automatic generation of forced checkpoints (somehow similarly to communication-induced checkpointing [35]), and the use of a reversible semantics. Also, we share some similarities with the checkpointing technique for fault-tolerant distributed computing of [8,17], although the aim is in principle different: their goal is the definition of a new programming model where globally consistent checkpoints can be created (rather than extending an existing message-passing programming language with explicit operators for rollback recovery). Indeed, the use of some form of reversibility is mentioned in [8] as future work.

The idea of using reversible computation for rollback recovery is not new. E.g., Perumalla and Park [33] already suggested it as an alternative to other traditional techniques based on checkpointing. In contrast to our work, the authors focus on empirically analyzing the trade-off between fault tolerance based on checkpointing and on reversible computation (i.e., memory vs run time), using a particular example (a particle collision application). Moreover, since the application is already reversible, no Landauer embedding is required.

The introduction of a rollback construct in a causal-consistent concurrent formalism can be traced back to [11,12,21,22,27]. In these works, however, the authors focus on a different formalism and, moreover, no explicit checkpointing operator is considered. These ideas are then transferred to an Erlang-like language in [23,31], where an explicit checkpoint operator is introduced. However, in contrast to our work, all actions are recorded into a history (i.e., it has no way of turning the reversible mode off). In other words, a checkpoint is just a mark in the execution, but it is not propagated to other processes (as our forced checkpoints) and cannot be removed (as a call to commit does in our approach). More recent formulations of the reversible semantics for an Erlang-like language include [20,24–26], but the checkpoint operator has not been considered (the focus is on reversible debugging).

The standard semantics in Fig. 1 is trivially equivalent to that considered in [25] except for some minor details: First, we follow the simpler and more elegant formulation of [26]. For instance, following the style of [25], rule *Send* would have the following form:

$$\frac{s \xrightarrow{\text{send}(p',v)} s'}{\Gamma; \langle p, s \rangle \mid \Pi \rightarrowtail_{p,\text{send}(\ell)} \Gamma \cup \{(p, p', v)\}; \langle p, s' \rangle \mid \Pi}$$

In this case, messages are stored in a global mailbox, Γ, and an expression like "$\langle p, s \rangle \mid \Pi$" represents all the processes in the system, i.e., $\langle p, s \rangle$ is a distinguished process (where reduction applies) and Π is the parallel composition of the remaining processes. In contrast, we have *floating* messages and select a process to be reduced by applying (repeatedly) rule *Par*. The possible reductions, though, are the same in both cases. There are other, minor differences, like considering a rule to deal with the predefined function *self* (which returns the pid of a process), and representing a state by a pair θ, e (environment, expression).

Another difference with the reversible semantics in [23, 25] is that we consider a single transition relation for systems ($\hookrightarrow$). This relation aims at modeling an actual execution in which a process proceeds normally forwards but a call to rollback forces it to go backwards temporarily (a situation that can be propagated to other processes in order to be causally consistent). In contrast, [23,25] considers first an *uncontrolled* semantics ($\rightharpoonup$ and $\leftharpoonup$) which models *all* possible forward and backward computations. Then, a *controlled* semantics is defined on top of it to drive the steps of the uncontrolled semantics in order to satisfy both replay and rollback requests.

On a different line of work, Vassor and Stefani [36] formally studied the relation between rollback recovery and causal-consistent reversible computation. In particular, they consider the relation between a distributed checkpoint/rollback scheme based on (causal) logging (Manetho [4]) and a causally-consistent reversible version of π-calculus with a rollback operator [21]. Their main conclusion is that the latter can simulate the rollback recovery strategy of Manetho. Our aim is somehow similar, since we also simulate a checkpoint-based rollback recovery strategy using a reversible semantics, but there are also some significant differences: the considered language is different (a variant of π-calculus vs an Erlang-like language), they only consider a fixed number of processes (while we accept dynamic process spawning) and, moreover, no explicit operators are considered (i.e., our approach is more oriented to introduce a new programming feature rather than proving a theoretical result).

Very recently, Mezzina, Tiezzi and Yoshida [29] introduced a rollback recovery strategy for session-based programming. Besides considering a different setting (a variant of π-calculus), their approach is also limited to a fixed number of parties (no dynamic processes can be added at run time), and nested checkpoints are not allowed. Furthermore, the checkpoints of [29] are not automatically propagated to other causally consistent processes (as our forced checkpoints); rather, they introduce a *compliance check* at the type level to prevent undesired situations.

Our work also shares some similarities with [34], which presents a hybrid model combining message-passing concurrency and software transactional memory. However, the underlying language is different and, moreover, their transactions cannot include process spawning (which must be delayed).

Finally, Fabbretti, Lanese and Stefani [6] introduced a calculus to formally model distributed systems subject to crash failures, where recovery mechanisms can be encoded by a small set of primitives. This work can be seen as a reworking and extension of the previous work by Francalanza and Hennessy [9]. Here, a variant of π-calculus is considered. Furthermore, the authors focus on crash recovery without relying on a form of checkpointing, in contrast to our approach.

5   Conclusions and Future Work

In this work, we have defined a rollback-recovery strategy for a message-passing concurrent programming language without the need for a central coordination. For this purpose, we have extended the underlying language with three explicit operators: check, commit, and rollback. Our approach is based on a reversible semantics where every process may go both forwards and backwards (during a rollback). Checkpoints are automatically propagated to other processes so that backward computations are causally consistent. The ability to turn the reversible mode on/off is useful not only to model rollback recovery, but can also constitute the basis of a safer try_catch (as illustrated in Sect. 1) and a *selective* reversible debugging scheme, where only some computations—those of interest—are traced, thus making it easier to scale to larger applications.

As for future work, we will consider the definition of a *shortcut* version of the rollback semantics where only the state in a checkpoint is recorded (rather than all the states between a checkpoint and the corresponding commit/rollback) so that a rollback recovers the saved state in one go. This extension will be essential to make our approach feasible in practice. In the context of Erlang, a prototype implementation of the proposed operators (check, commit, and rollback) could be carried out through a program instrumentation. It will likely require introducing a *wrapper* for each process in order to record the process' history, turning the reversible mode on/off, propagating forced checkpoints and commits, etc. For this purpose, one could explore the use of the run-time monitors of [10], which play a similar role in their scheme for reversible choreographies.

Acknowledgements. The author would like to thank Ivan Lanese and Adrián Palacios for their useful remarks and discussions on a preliminary version of this work. I would also like to thank the anonymous reviewers and the participants of FACS 2023 for their suggestions to improve this paper.

References

1. Aman, B., et al.: Foundations of reversible computation. In: Ulidowski, I., Lanese, I., Schultz, U.P., Ferreira, C. (eds.) Reversible Computation: Extending Horizons of Computing - Selected Results of the COST Action IC1405. LNCS, vol. 12070, pp. 1–40. Springer, Cham (2020). https://doi.org/10.1007/978-3-030-47361-7_1
2. Danos, V., Krivine, J.: Reversible communicating systems. In: Gardner, P., Yoshida, N. (eds.) CONCUR 2004. LNCS, vol. 3170, pp. 292–307. Springer, Heidelberg (2004). https://doi.org/10.1007/978-3-540-28644-8_19
3. Elnozahy, E.N., Alvisi, L., Wang, Y., Johnson, D.B.: A survey of rollback-recovery protocols in message-passing systems. ACM Comput. Surv. **34**(3), 375–408 (2002)
4. Elnozahy, E.N., Zwaenepoel, W.: Manetho: transparent rollback-recovery with low overhead, limited rollback, and fast output commit. IEEE Trans. Comput. **41**(5), 526–531 (1992). https://doi.org/10.1109/12.142678
5. Erlang website (2021). https://www.erlang.org/
6. Fabbretti, G., Lanese, I., Stefani, J.B.: A behavioral theory for crash failures and erlang-style recoveries in distributed systems. Technical report. RR-9511, INRIA (2023). https://hal.science/hal-04123758
7. Frequently Asked Questions about Erlang (2018). http://erlang.org/faq/academic.html
8. Field, J., Varela, C.A.: Transactors: a programming model for maintaining globally consistent distributed state in unreliable environments. In: Palsberg, J., Abadi, M. (eds.) Proceedings of the 32nd ACM SIGPLAN-SIGACT Symposium on Principles of Programming Languages (POPL 2005), pp. 195–208. ACM (2005)
9. Francalanza, A., Hennessy, M.: A theory of system behaviour in the presence of node and link failure. Inf. Comput. **206**(6), 711–759 (2008). https://doi.org/10.1016/j.ic.2007.12.002
10. Francalanza, A., Mezzina, C.A., Tuosto, E.: Reversible choreographies via monitoring in Erlang. In: Bonomi, S., Rivière, E. (eds.) Proceedings of the 18th IFIP WG 6.1 International Conference on Distributed Applications and Interoperable Systems (DAIS 2018), Held as Part of DisCoTec 2018. LNCS, vol. 10853, pp. 75–92. Springer, Cham (2018). https://doi.org/10.1007/978-3-319-93767-0_6
11. Giachino, E., Lanese, I., Mezzina, C.A.: Causal-consistent reversible debugging. In: Gnesi, S., Rensink, A. (eds.) FASE 2014. LNCS, vol. 8411, pp. 370–384. Springer, Heidelberg (2014). https://doi.org/10.1007/978-3-642-54804-8_26
12. Giachino, E., Lanese, I., Mezzina, C.A., Tiezzi, F.: Causal-consistent reversibility in a tuple-based language. In: Daneshtalab, M., Aldinucci, M., Leppänen, V., Lilius, J., Brorsson, M. (eds.) Proceedings of the 23rd Euromicro International Conference on Parallel, Distributed, and Network-Based Processing, PDP 2015, pp. 467–475. IEEE Computer Society (2015)
13. Glück, R., et al.: Towards a taxonomy for reversible computation approaches. In: Kutrib, M., Meyer, U. (eds.) Reversible Computation, pp. 24–39. Springer, Cham (2023). https://doi.org/10.1007/978-3-031-38100-3_3
14. González-Abril, J.J., Vidal, G.: Causal-consistent reversible debugging: improving CauDEr. Technical report, DSIC, Universitat Politècnica de València (2020). https://gvidal.webs.upv.es/confs/padl21/tr.pdf
15. González-Abril, J.J., Vidal, G.: Causal-consistent reversible debugging: improving CauDEr. In: Morales, J.F., Orchard, D.A. (eds.) Proceedings of the 23rd International Symposium on Practical Aspects of Declarative Languages (PADL 2021). LNCS, vol. 12548, pp. 145–160. Springer, Cham (2021). https://doi.org/10.1007/978-3-030-67438-0_9

16. Hewitt, C., Bishop, P.B., Steiger, R.: A universal modular ACTOR formalism for artificial intelligence. In: Nilsson, N.J. (ed.) Proceedings of the 3rd International Joint Conference on Artificial Intelligence, pp. 235–245. William Kaufmann (1973). http://ijcai.org/Proceedings/73/Papers/027B.pdf

17. Kuang, P., Field, J., Varela, C.A.: Fault tolerant distributed computing using asynchronous local checkpointing. In: Boix, E.G., Haller, P., Ricci, A., Varela, C. (eds.) Proceedings of the 4th International Workshop on Programming Based on Actors Agents & Decentralized Control (AGERE! 2014), pp. 81–93. ACM (2014)

18. Lamport, L.: Time, clocks, and the ordering of events in a distributed system. Commun. ACM **21**(7), 558–565 (1978). https://doi.org/10.1145/359545.359563

19. Landauer, R.: Irreversibility and heat generation in the computing process. IBM J. Res. Dev. **5**, 183–191 (1961)

20. Lanese, I., Medic, D.: A general approach to derive uncontrolled reversible semantics. In: Konnov, I., Kovács, L. (eds.) 31st International Conference on Concurrency Theory, CONCUR 2020. LIPIcs, vol. 171, pp. 33:1–33:24. Schloss Dagstuhl - Leibniz-Zentrum für Informatik (2020). https://doi.org/10.4230/LIPIcs.CONCUR.2020.33

21. Lanese, I., Mezzina, C.A., Schmitt, A., Stefani, J.-B.: Controlling reversibility in Higher-Order Pi. In: Katoen, J.-P., König, B. (eds.) CONCUR 2011. LNCS, vol. 6901, pp. 297–311. Springer, Heidelberg (2011). https://doi.org/10.1007/978-3-642-23217-6_20

22. Lanese, I., Mezzina, C.A., Stefani, J.: Reversibility in the higher-order π-calculus. Theor. Comput. Sci. **625**, 25–84 (2016)

23. Lanese, I., Nishida, N., Palacios, A., Vidal, G.: A theory of reversibility for Erlang. J. Log. Algebraic Methods Program. **100**, 71–97 (2018). https://doi.org/10.1016/j.jlamp.2018.06.004

24. Lanese, I., Palacios, A., Vidal, G.: Causal-consistent replay debugging for message passing programs. In: Pérez, J.A., Yoshida, N. (eds.) Proceedings of the 39th IFIP WG 6.1 International Conference on Formal Techniques for Distributed Objects, Components, and Systems (FORTE 2019). LNCS, vol. 11535, pp. 167–184. Springer, Cham (2019). https://doi.org/10.1007/978-3-030-21759-4_10

25. Lanese, I., Palacios, A., Vidal, G.: Causal-consistent replay reversible semantics for message passing concurrent programs. Fundam. Informaticae **178**(3), 229–266 (2021). https://doi.org/10.3233/FI-2021-2005

26. Lanese, I., Sangiorgi, D., Zavattaro, G.: Playing with bisimulation in Erlang. In: Boreale, M., Corradini, F., Loreti, M., Pugliese, R. (eds.) Models, Languages, and Tools for Concurrent and Distributed Programming – Essays Dedicated to Rocco De Nicola on the Occasion of His 65th Birthday. LNCS, vol. 11665, pp. 71–91. Springer, Cham (2019). https://doi.org/10.1007/978-3-030-21485-2_6

27. Lienhardt, M., Lanese, I., Mezzina, C.A., Stefani, J.B.: A reversible abstract machine and its space overhead. In: Giese, H., Rosu, G. (eds.) Proceedings of the Joint 14th IFIP WG International Conference on Formal Techniques for Distributed Systems (FMOODS 2012) and the 32nd IFIP WG 6.1 International Conference (FORTE 2012). LNCS, vol. 7273, pp. 1–17. Springer, Cham (2012). https://doi.org/10.1007/978-3-642-30793-5_1

28. Matsuda, K., Hu, Z., Nakano, K., Hamana, M., Takeichi, M.: Bidirectionalization transformation based on automatic derivation of view complement functions. In: Hinze, R., Ramsey, N. (eds.) Proceedings of the 12th ACM SIGPLAN International Conference on Functional Programming, ICFP 2007, pp. 47–58. ACM (2007)

29. Mezzina, C.A., Tiezzi, F., Yoshida, N.: Rollback recovery in session-based programming. In: Jongmans, S., Lopes, A. (eds.) Proceedings of the 25th IFIP WG 6.1 International Conference on Coordination Models and Languages, COORDINATION 2023. LNCS, vol. 13908, pp. 195–213. Springer, Cham (2023). https://doi.org/10.1007/978-3-031-35361-1_11

30. Milner, R. (ed.): A Calculus of Communicating Systems. LNCS, vol. 92. Springer, Heidelberg (1980). https://doi.org/10.1007/3-540-10235-3

31. Nishida, N., Palacios, A., Vidal, G.: A reversible semantics for Erlang. In: Hermenegildo, M., López-García, P. (eds.) Proceedings of the 26th International Symposium on Logic-Based Program Synthesis and Transformation (LOPSTR 2016). LNCS, vol. 10184, pp. 259–274. Springer, Cham (2017). https://doi.org/10.1007/978-3-319-63139-4_15

32. Nishida, N., Palacios, A., Vidal, G.: Reversible computation in term rewriting. J. Log. Algebraic Methods Program. **94**, 128–149 (2018). https://doi.org/10.1016/j.jlamp.2017.10.003

33. Perumalla, K.S., Park, A.J.: Reverse computation for rollback-based fault tolerance in large parallel systems - evaluating the potential gains and systems effects. Clust. Comput. **17**(2), 303–313 (2014). https://doi.org/10.1007/s10586-013-0277-4

34. Swalens, J., Koster, J.D., Meuter, W.D.: Transactional actors: communication in transactions. In: Jannesari, A., de Oliveira Castro, P., Sato, Y., Mattson, T. (eds.) Proceedings of the 4th ACM SIGPLAN International Workshop on Software Engineering for Parallel Systems, SEPSSPLASH 2017, pp. 31–41. ACM (2017). https://doi.org/10.1145/3141865.3141866

35. Tsai, J., Wang, Y.: Communication-induced checkpointing protocols and rollback-dependency trackability: a survey. In: Wah, B.W. (ed.) Wiley Encyclopedia of Computer Science and Engineering. Wiley (2008). https://doi.org/10.1002/9780470050118.ecse059

36. Vassor, M., Stefani, J.B.: Checkpoint/Rollback vs Causally-consistent reversibility. In: Kari, J., Ulidowski, I. (eds.) Reversible Computation, pp. 286–303. Springer, Cham (2018). 978-3-319-99498-7_20, https://doi.org/10.1007/978-3-319-99498-7_20

37. Vidal, G.: From reversible computation to checkpoint-based rollback recovery for message-passing concurrent programs. CoRR abs/2309.04873 (2023). https://arxiv.org/abs/2309.04873

Anniversary Papers

Formal Model Engineering of Distributed CPSs Using AADL: From Behavioral AADL Models to Multirate Hybrid Synchronous AADL

Kyungmin Bae[1]([✉]) and Peter Csaba Ölveczky[2]

[1] Pohang University of Science and Technology, Pohang, South Korea
kmbae@postech.ac.kr
[2] University of Oslo, Oslo, Norway

Abstract. A promising way of integrating formal methods into industrial system design is to endow industrial modeling tools with automatic formal analyses. In this paper we identify some challenges for providing such formal methods "backends" for cyber-physical systems (CPSs), and argue that Maude could meet these challenges. We then give an overview of our research on integrating Maude analysis into the OSATE tool environment for the industrial CPS modeling standard AADL.

Since many critical distributed CPSs are "logically synchronous," a key feature making automatic formal analysis practical is the use of *synchronizers* for CPSs. We identify a sublanguage of AADL to describe *synchronous* CPS designs. We can then use Maude to effectively verify such synchronous designs, which under certain conditions also verifies the corresponding asynchronous distributed systems, with clock skews and communication delays. We then explain how we have extended our methods to *multirate* systems and to CPSs with continuous behaviors.

We illustrate the effectiveness of Maude-based formal model engineering of industrial CPSs on avionics control systems and collections of drones. Finally, we identify future directions in this line of research.

1 Introduction and Overview

Modern cyber-physical systems (CPSs) are complex and safety-critical systems. Formal methods should therefore be an integral part of their design process. However, despite the availability of powerful formal tools, there still seems to be a barrier in industry to using formal methods. This could have many reasons, including lack of available formal methods experts, perceived need to be an expert to use formal methods, and lack of user-friendly and robust formal tools.

Formal Model Engineering. Maybe the most promising way of integrating formal methods into industrial model development processes is to provide automatic "push-button" formal analysis as an "under-the-hood backend" which is integrated into the designer's model development environment. The modeler can then continue to use her favorite modeling language and IDE without worrying

J. Cámara and S.-S. Jongmans (Eds.): FACS 2023, LNCS 14485, pp. 127–152, 2024.
https://doi.org/10.1007/978-3-031-52183-6_7

about formal methods, and still get powerful formal analyses *for free*. Edward Lee calls such model development *formal model engineering* in [8].

Enabling such formal model engineering requires:

- Targeting a modeling language widely used in industrial model development.
- Identifying a useful significant subset of the industrial modeling language.
- A formal semantics for the (subset of the) industrial modeling language.
- Defining an intuitive property specification language in which the developer can easily express properties about her model, without knowing anything about formal methods or the formal representation of her model, and without having to master the syntax of the formal method.
- Integrating *automatic* formal analyses of the user's model into *her* tool.
- Presenting the result of the formal analysis in an intuitive way.

Challenges. Endowing industrial modeling environments for CPSs, which are distributed (real-time) embedded systems, with effective under-the-hood formal analysis is very challenging, due to factors that include:

1. Industrial modeling languages by their very nature tend to be large and expressive. The target formalism must therefore be very expressive to provide a formal semantics for the source language.
2. However, in formal methods usually only less expressive, often decidable, formalisms (such as timed automata for real-time systems) provide *automatic* analyses, whereas more expressive formalisms (such as differential dynamic logic [61]) typically only support interactive theorem proving analysis, which is not desired for formal model engineering.
3. Automatic model checking verification of industrial distributed CPSs quickly becomes infeasible due to the many interleavings caused by distribution.
4. Industrial CPSs may have both *continuous* behaviors and complex *discrete* control programs, in addition to clock skews, network delays, and so on.
5. The need to provide a query language in which requirements can be intuitively expressed without knowing the formal representation of the model.
6. Verification typically applies to a specific deployment scenario, and small changes (such as network delays) can invalidate the verification result.

Maude as a Formal Model Engineering Backend for CPSs. We argue that Maude [27], with its Real-Time Maude [58,59] and Maude-SMT [72] extensions, is a suitable formal framework for formal model engineering of industrial CPSs.

Maude is formal modeling language and high-performance formal analysis tool based on rewriting logic [48,50]. Rewriting logic is an expressive logic for concurrency where the static parts of the system (data types, etc.) are defined by algebraic equational specifications, and where dynamic behavior is defined by rewrite rules. Maude is particularly suitable to model distributed systems in an object-oriented style. Maude provides a range of explicit-state formal analysis methods, including rewriting for simulation, reachability analysis, and linear temporal logic (LTL) model checking. Real-time systems can also be specified and analyzed in Real-Time Maude, which provides time-bounded analysis commands and timed CTL model checking [44]. Maude has recently integrated SMT solving, which enables *symbolic* reasoning that manipulates *symbolic* states; i.e., state patterns with variables that represent possibly infinitely many concrete states.

Maude addresses the challenges 1, 2, 4, and 5 above as follows:

1. Rewriting logic is well known to be expressive (see, e.g., [50] for a dated overview of some applications), and can capture complex control program languages—evidenced, e.g., by rewriting logic providing semantics to programming languages like C, Java, LLVM, EVM, Scheme, and so on [53]—and a wide range of communication forms, including sophisticated communication models for wireless sensor networks [60] and mobile ad hoc networks [46].
2. Maude and Real-Time Maude combine this expressive and general modeling formalism with providing a range of *automatic* analysis methods, including reachability analysis and LTL and timed CTL model checking.
3. The recent integration of SMT solving into Maude allows us to analyze systems with *continuous* behaviors (and clock skews) symbolically in Maude.
4. In contrast to many formal tools, Maude supports *parametric* atomic propositions, which are needed to define intuitive property specification languages (which are parametric in the input model), as illustrated below.

(Logically) Synchronous CPSs. We target CPSs, which, by definition, are *networks* of embedded systems that typically interact using message passing. Many industrial (distributed) CPSs are *logically synchronous*: at the end of each period all components read messages, perform transitions, and generate output to be read at the end of the next period. Examples of such logically synchronous systems include avionics and automotive systems [9,43,67], networked medical devices [7,36], and other distributed control systems such as the steam-boiler benchmark [1]. However, they have to be realized in a distributed setting, with imprecise local clocks, messaging delays, execution times, and so on.

Note that clock synchronization is well understood (e.g., IEEE 1588 [30]); we can therefore often guarantee a bound on the drift of the local clocks. Furthermore, the infrastructure of many critical CPSs—such as cars, airplanes, manufacturing plants, robots, etc.—guarantee bounds on the communication delays.

Synchronizers for CPSs: Taming the Design and Model Checking Complexity. Challenge (3) still remains: model checking distributed CPSs quickly becomes infeasible due to the many different behaviors caused by interleavings.

One way of making model checking of logically synchronous distributed CPSs feasible is to apply *synchronizers for CPSs*, such as the *time-triggered architecture* (TTA) [38,65], PALS ("physically asynchronous, logically synchronous") [3,52], and their generalization MSYNC [13]. The idea is that an idealized *fully synchronous design SD* (*without* distribution, message delays, clock skews, execution times, etc.) under some assumptions Γ about the underlying infrastructure is "logically equivalent" to the desired asynchronous distributed system $async(SD, \Gamma)$.

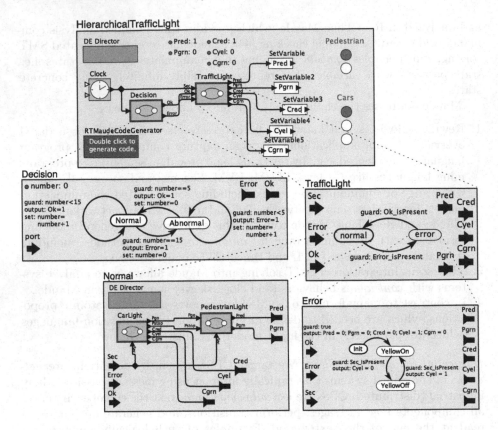

Fig. 1. A Ptolemy II DE model of a traffic light system.

Synchronizers for CPSs also address Challenge (6): By verifying the synchronous design, we also prove the correctness of the corresponding "implementations" for *all* deployments satisfying the TTA/PALS/MSYNC constraint.

PALS was motivated by an avionics system developed by Rockwell Collins, where model checking even very simplified versions of the (asynchronous) CPS design took more than 30 h, whereas model checking the logically equivalent synchronous design takes less than 0.1 s.

AADL. In [15] we provided a Real-Time Maude formal analysis backend for *discrete-event* (DE) models of UC Berkeley's Ptolemy II tool [62], which has industrial users (e.g., Bosch). As shown in Fig. 1, the user can develop her model using the excellent graphical interface of Ptolemy II. She can then click on the blue button to generate a Real-Time Maude model and can enter her Real-Time Maude query, all while staying inside Ptolemy II. Such Real-Time Maude timed CTL model checking revealed a previously unknown flaw in a model of a traffic light system in Ptolemy II's model repository [44] (Fig. 2).

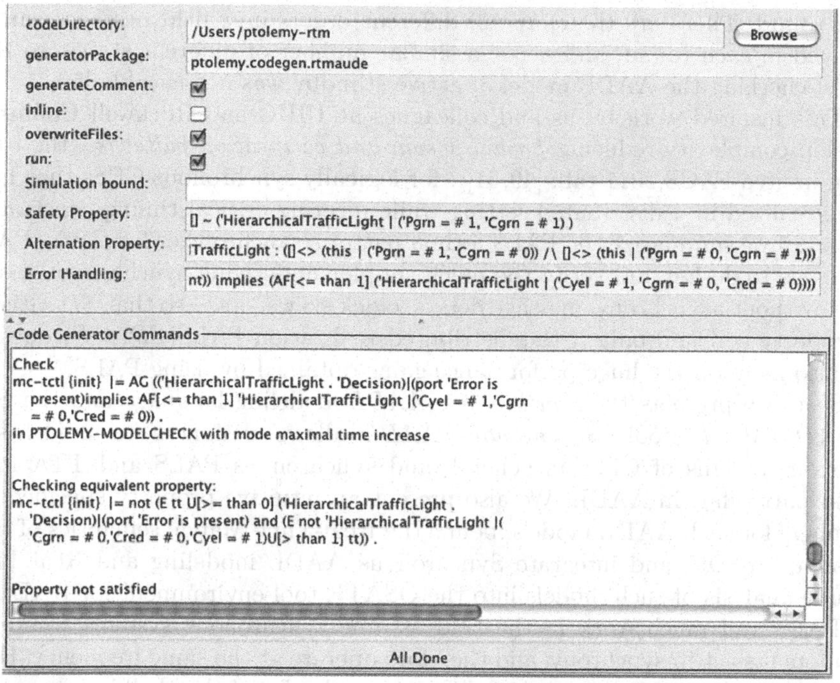

Fig. 2. Real-Time Maude timed CTL model checking inside Ptolemy II.

We wanted to target industrial tools widely used by designers of safety-critical CPSs. In particular, due to our collaboration with Rockwell Collins on avionics applications, we targeted a modeling standard for CPSs used in avionics. The *Architecture Analysis and Design Language* (AADL) [31] is an industrial modeling standard for avionics, automotive, and cyber-physical systems developed and used by companies and organizations such as Carnegie Mellon University, US Army, Honeywell, Rockwell Collins, Lockheed Martin, General Dynamics, Airbus, the European Space Agency, Dassault, EADS, Ford, and Toyota. Model development in AADL is supported by the Open Source AADL Tool Environment (OSATE).

Overview. This invited paper on "the evolution of our software-component-based" research first provides brief preliminaries to Maude and its extensions, and to AADL. Section 3 introduces two avionics applications that motivated this work.

Section 4 summarizes the first step towards providing Maude-based formal analysis of CPSs via AADL: a Real-Time Maude semantics of a "behavioral subset" of AADL [55]. This can be used to verify AADL models of distributed systems. However, explicit-state model checking of distributed CPSs quickly becomes infeasible: The above-mentioned *active standby* avionics system has only three components, and 10 boolean-valued messages are sent/received in each round. There are therefore 10! different orders in which messages can be received in each round, and hence a similar number of different states, so that model checking the AADL model of active standby was not feasible [56].

This inspired work by us and colleagues at UIUC and Rockwell Collins to develop complexity-reducing *formal design and verification patterns*—the topic of an invited FACS 2011 talk [49,51]—for logically synchronous CPSs that have to be realized in a distributed setting while meeting critical timing constraints. In Sect. 5 we summarize the PALS formal pattern/synchronizer for CPSs. PALS allows us to design and verify the much simpler underlying synchronous design *SD*—without asynchrony, message delays, clock skews, etc.—so that *SD* satisfies a property ϕ if and only if the distributed realization PALS(SD, p, Γ) does so. We also mention the huge performance gains obtained by using PALS.

Section 6 presents the *Synchronous AADL* modeling language and the *SynchAADL2Maude* tool. *Synchronous AADL* allows designers to model synchronous designs of CPSs in general, and synchronous PALS and TTA models in particular, in AADL. We also present an intuitive property specification language for such AADL models, define the Real-Time Maude semantics of Synchronous AADL, and integrate Synchronous AADL modeling and Real-Time Maude analysis of such models into the OSATE tool environment for AADL.

PALS and much work in the field assume synchronous systems where all components act in synchrony, and therefore operate at the same frequency. However, some CPSs are composed of different kinds of "off-the-shelf" components which operate at *different* frequencies. One prototypical example is an aircraft: The aileron controllers of commercial aircrafts typically operate at frequency 30–100 Hz, whereas the rudder controller operates at 30–50 Hz [2], yet they (and other controllers) need to synchronize to make a safe turn. With José Meseguer we therefore extended PALS (in a FACS 2012 paper [11]), Synchronous AADL, and the SynchAADL2Maude tool to the *multi-rate* setting. In one application, we use Euler approximations of continuous dynamics to analyze a textbook algorithm for turning an airplane. This work is summarized in Sect. 7.

The airplane turning example emphasizes that many CPSs have *continuous* environments. Defining synchronizers in this case is tricky, since we can no longer abstract from the exact time when a continuous environment is sampled and actuated; times which depend on the imprecise local clocks. To capture all possible behaviors depending on imprecise local clocks, and continuous environments, we combine Maude analysis with SMT solving. We summarize this *Hybrid PALS* synchronizer and the Maude+SMT-based formal analysis of logically synchronous *hybrid* CPSs that we have integrated into OSATE in Sect. 8.

Finally, we discuss future directions in this line of research in Sect. 9, and give some concluding remarks in Sect. 10.

2   Preliminaries

AADL. The *Architecture Analysis & Design Language* (AADL) [31] is an industrial modeling standard used in avionics, aerospace, automotive, medical devices, and robotics to describe an embedded real-time system as an assembly of software components mapped onto a hardware platform. A component *type* specifies the component's *interface* (e.g., ports) and *properties* (e.g., periods), and a component *implementation* specifies its internal structure as a set of *subcomponents* and a set of *connections* linking their ports. An AADL construct may have *properties* describing its parameters, declared in *property sets*. The OSATE modeling environment provides a set of Eclipse plug-ins for AADL.

Software components include *threads* that model the application software and *data* components representing data types. *System* components are the top-level components. A port is a *data* port, an *event* port, or an *event data* port. A component can have different *modes* and mode-specific property values, subcomponents, etc. Mode transitions are triggered by events.

Thread behavior is modeled as a guarded transition system with local variables using AADL's *Behavior Annex* [32]. A *periodic* thread is activated at fixed time intervals, and an *aperiodic* thread is activated when it receives an event. When a thread is activated, transitions are applied until a *complete* state is reached (or the thread suspends). The actions performed when a transition is applied may update local variables, generate outputs, and/or suspend the thread. Actions are built from basic actions using sequencing, conditionals, and finite loops.

Maude and Real-Time Maude. Maude [27] is a an executable formal specification language and high-performance analysis tool for distributed systems. A Maude module specifies a *rewrite theory* [48] $\mathcal{R} = (\Sigma, E, L, R)$, where:

- Σ is an algebraic *signature*, i.e., a set of *sorts*, *subsorts*, and *function symbols*.
- (Σ, E) is an *order-sorted equational logic theory* [35] specifying the system's data types, with E a set of (possibly conditional) equations and axioms.
- L is a set of rule *labels*.
- R is a collection of *labeled conditional rewrite rules* **crl** $[l] : t \Rightarrow t'$ **if** *cond*, with t, t' Σ-terms and $l \in L$, that specify the system's local transitions.

A one-step rewrite $t \longrightarrow_{\mathcal{R}} t'$ [48] holds iff t can be rewritten to t' by a rewrite rule in $\mathcal{R}$, and $\longrightarrow_{\mathcal{R}}^*$ denotes the reflexive-transitive closure of $\longrightarrow_{\mathcal{R}}$.

A declaration **class** $C \mid att_1 : s_1, \ldots, att_n : s_n$ declares an *object class* C with attributes att_1 to att_n of sorts s_1 to s_n. An *object* of class C is a term $<o : C \mid att_1 : val_1, \ldots, att_n : val_n>$, where o (of sort Oid) is the object's *identifier*, and val_1 to val_n are the current values of the attributes att_1 to att_n. *Messages* are terms of sort Msg. A system state is modeled as a term of sort Configuration, and consists of a *multiset* of objects and messages. The system's transitions are specified using rewrite rules.

The rewrite command simulates one behavior of the system. The command search [n] $t_0 \Rightarrow* pattern$ such that *cond* searches for at most n states reachable from state t_0 that match *pattern* and satisfy the condition *cond*. Maude's linear temporal logic (LTL) model checker checks whether all paths from the initial state satisfy an LTL formula. Such formulas are constructed by (possibly parametric) state propositions of sort Prop, and the usual LTL operators True, ~ (negation), \/, /\, ->, [] ("always"), <> ("eventually"), U ("until"), and O ("next").

To specify real-time systems, Real-Time Maude [57,59] adds *tick* rewrite rules crl [*l*] : {t_1} => {t_2} in time τ if *cond* to model *time elapse*, where the whole state has the form {t}. This rule specifies that it takes time $\tau\sigma$ for the state {$t_1\sigma$} to evolve to the state {$t_2\sigma$}, provided that the condition *cond* holds for the matching substitution σ. Since the whole state has the form {t}, and {_} cannot appear in a subterm of t, the form of the tick rewrite rules ensures that time elapses in the *entire* state when a tick rule is applied. Other (non-tick) rewrite rules are considered to take zero time. Real-Time Maude provides a range of time-bounded and unbounded search, LTL, and timed CTL model checking commands [44,57]; the time-bounded LTL model checking command is written (mc t_0 |=t *formula* in time <= τ).

Maude+SMT. *Constrained terms* [20,63] symbolically represent (possibly infinite) sets of system states. A constrained term is a pair $\phi \parallel t$ of a constraint $\phi(x_1,\ldots,x_n)$ and a term $t(x_1,\ldots,x_n)$ over SMT variables $x_1,\ldots,x_n$. It represents the set $[\![\phi \parallel t]\!]$ of all instances of the pattern t such that ϕ holds.

A one-step *symbolic rewrite* $\phi_t \parallel t \rightsquigarrow_{\mathcal{R}} \phi_u \parallel u$ on constrained terms [63] symbolically represents a (possibly infinite) set of system transitions. We denote by $\rightsquigarrow_{\mathcal{R}}^*$ the reflexive-transitive closure of $\rightsquigarrow_{\mathcal{R}}$. For a symbolic rewrite $\phi_t \parallel t \rightsquigarrow_{\mathcal{R}}^* \phi_u \parallel u$, there exists a "concrete" rewrite $t' \longrightarrow_{\mathcal{R}}^* u'$ with $t' \in [\![\phi_t \parallel t]\!]$ and $u' \in [\![\phi_u \parallel u]\!]$. Conversely, for any concrete rewrite $t' \longrightarrow_{\mathcal{R}}^* u'$ with $t' \in [\![\phi_t \parallel t]\!]$, there exists a symbolic rewrite $\phi_t \parallel t \rightsquigarrow_{\mathcal{R}}^* \phi_u \parallel u$ with $u' \in [\![\phi_u \parallel u]\!]$.

Maude provides SMT solving and *symbolic reachability analysis* for constrained terms, using connections to Yices2 [28] and CVC4 [21]. Maude supports SMT theories for Booleans, integers, and reals in the SMT-LIB standard [22].

3   Two Motivating Applications

Even though the methods summarized in this paper are applicable to many CPSs, much of the work was motivated by two (classes of) avionics applications:

1. An avionics control system developed at Rockwell Collins whose modeling and, in particular, NuSMV model checking was much harder than expected.
2. Together with a student, Joshua Krisiloff, at the UIUC aeronautics department we wanted to investigate to what degree formal methods could be used to analyze control algorithms for airplanes, for example for turning an airplane.

The Active Standby System. There are multiple physically separated *cabinets* ("main computer systems") on an aircraft, so that physical damage does not take out the computer system. The *active standby* system [54] developed at Rockwell Collins focuses on the logic for deciding which of *two* cabinets is *active*.

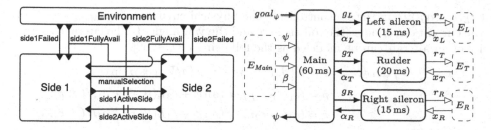

Fig. 3. The active standby system.   **Fig. 4.** Turning an airplane.

The two sides/cabinets receive inputs through communication channels. Each side can fail, and recover after failure. If one side fails, the non-failed side should be the active side. The pilot can also toggle the active status of these sides. The *full* functionality of each side is dependent on the two sides' perception of the availability of other system components. The architecture of the system is shown in Fig. 3. The system consists of three components: Side 1, Side 2, and Environment. The Environment component is an abstract representation of other components that interact with Side 1 and Side 2. The components have the same period and dispatch at the same time. Each time Environment dispatches, it sends five Boolean (nondeterministically generated) values, denoting, respectively, whether side i is failed or not fully available, and whether the pilot wants to change the active side ("manualSelection"). The connections between the two sides are "delayed;" a message sent in one round is read in the next round.

There are five important properties that active standby should satisfy [52, 54]. One of them is: *A side that is not fully available should not be the active side if the other side is fully available (provided neither side has failed, the availability of a side has not changed, and the pilot has not made a manual selection).*

Turning an Airplane. An airplane turns by *rolling* in the direction of the turn by moving its two *ailerons* (flaps attached to each wing), and reduces the *adverse yaw* caused by the rolling by moving its *rudder* (flap attached to the vertical stabilizer). To achieve a smooth turn these devices should synchronize, even though their controllers have different periods.

A controller orchestrates these devices to turn an aircraft [9, 16]. As shown in Fig. 4, there are four components. The main controller has period 60 ms, the aileron subcontrollers have period 15 ms, and the rudder subcontroller has period 20 ms. Each component interacts with its *physical* environment.

Each subcontroller $M \in \{L, T, R\}$ determines the moving rate r_M to move its surface towards the goal angle g_M provided by the main controller. It also sends the "sampled" angle α_M to the main controller. Its physical environment E_M specifies the continuous behavior of the surface angle x_M by the control command r_M. The angle x_M gradually changes according to the ODE: $\dot{x}_M = r_M$.

The main controller determines the goal angles for the subcontrollers to make a coordinated turn, given a goal direction $goal_\psi$ and the sampled surface angles

$(\alpha_L, \alpha_T, \alpha_R)$ from the subcontrollers. The physical environment E_{Main} specifies the current direction ψ, the roll angle ϕ, and the yaw angle β. The lateral dynamics of an aircraft can be modeled as the following ODEs [69]:

$$\dot{\psi} = (g/V)\tan\phi, \qquad \dot{\beta} = Y(\beta, \vec{x})/mV - r + (g/V)\cos\beta\sin\phi,$$

$$\dot{\phi} = p \qquad\qquad \dot{p} = (c_1 r + c_2 p)\cdot r\tan\phi + c_3 L(\beta, \vec{x}) + c_4 N(\beta, \vec{x}),$$

$$\dot{r} = (c_8 p - c_2 r)\cdot r\tan\phi + c_4 L(\beta, \vec{x}) + c_9 N(\beta, \vec{x}),$$

where g is the gravitational constant, m is the mass of the aircraft, V is the velocity of the aircraft, p is the rolling moment, r is the yawing moment, and Y, L, and N are linear functions of β and the surface angles $\vec{x} = (x_L, x_V, x_R)$.

The yaw angle should be close to 0 during a turn, and the airplane should reach the goal direction with both roll angle and yaw angle close to 0 [9].

4 Formal Semantics and Analysis for "Behavioral AADL"

Although AADL is an industrial (SAE, Society of Automotive Engineers) standard modeling language for safety-critical embedded systems, it lacks a formal semantics. In [55], with colleagues at UIUC and Leicester, we therefore defined, for the first time (see [55] for a discussion on related work) an executable formal semantics for what we call a "behavioral subset" of AADL, in Real-Time Maude.

AADL is a huge and complex standard. Formal approaches therefore target limited fragments of AADL. We targeted a behavioral subset of AADL suitable to define distributed software designs—with all software structuring mechanisms of AADL (system, process, and thread components); (event, event data, and data) ports and their connections; mode-specific properties; mode transitions; and both periodic, aperiodic, sporadic, and background thread dispatch; and so on. We did not target the "hardware platform" part of AADL. Thread behaviors are modeled using AADL's Behavior Annex standard [32].

In [55] we defined the Real-Time Maude semantics for this subset of AADL. AADL components are *hierarchical*; the formal model should reflect the hierarchical structure, e.g., for understanding or "mapping back" analysis results.

Some key Real-Time Maude features that are crucial to define a decently effective semantics for this fragment of AADL include:

- An expressive formalism is needed to capture this rich subset of AADL, including a Turing-complete programming language used to define transitions.
- *Hierarchical objects* (an object attribute may have sort `Configuration` and therefore contain subconfigurations) allow us to define the semantics in an object-based style, yet preserve the hierarchical structure of AADL models.
- Rewriting logic's division into equations and rewrite rules—where only the latter contribute to the state space—is crucial to provide an efficient semantics. For example, "executing" the program associated with each AADL transition can be done equationally (i.e., in one atomic step).

Real-Time Maude Semantics. We can only give a very brief sample of our semantics, and refer to [56] for details. The key observation is that the semantics of a component-based language naturally can be defined in an object-oriented style, where each component instance is modeled as an object. As mentioned, the hierarchical structure of AADL components is reflected in the nested structure of objects. Any AADL component instance is represented as an object instance of a subclass of the following class Component, which contains the attributes common to all kinds of components (systems, processes, threads, etc.):

```
class Component | features : Configuration,    subcomponents : Configuration,
                  properties : Properties,    connections : ConnectionSet,
                  modes : Modes,              inModes : ModeNameSet .
```

The attribute features denotes the features of a component (i.e., its ports); subcomponents denotes the subcomponents of the object; properties denotes its *properties*, such as the dispatch protocol for threads; connections denotes the set of port connections of the object; modes contains the object's mode transition system; and inModes gives the set of modes (of the immediate supercomponent) in which the component is available. The Thread class is declared as follows:

```
class Thread | behavior : ThreadBehavior,    status : ThreadStatus,
               deactivated : Bool .
subclass Thread < Component .
```

The behavior attribute denotes the transition system associated with the thread. The status indicates the current status of the thread (active, completed, suspended, etc.). The attribute deactivated indicates whether the thread is deactivated because it is not in the current "active" modes of the system.

The following rewrite rule specifies the execution of an *active* thread. If the thread is in state L1, a transition from L1 whose guard evaluates to true is executed. The resulting status is sleeping(...) if the statement list SL contains delay statements; otherwise, the thread is completed or inactive if the resulting state L2 is a complete state, and remains active if not:

```
crl [apply-transition] :
< O : Thread | status : active,  deactivated : B,  features : PORTS,
               behavior : states current: L1 complete: LS1 others: LS2
                          state variables VAL
                          transitions (L1 -[GUARD]-> L2 {SL}) ; TRANSITIONS >
=>
< O : Thread | status : (if SLEEP-TIME > 0 then sleeping(SLEEP-TIME) else
                          (if (not L2 in LS1) then active else
                            (if B then inactive else completed fi) fi) fi),
               features : NEW-PORTS,
               behavior : states current: L2 complete: LS1 others: LS2
                          state variables NEW-VAL
                          transitions (L1 -[GUARD]-> L2 {SL}) ; TRANSITIONS >
  if evalGuard(GUARD, PORTS, VAL)
  /\ transResult(NEW-PORTS, NEW-VAL, SLEEP-TIME) :=
          executeTransition(L1 -[GUARD]-> L2 {SL}, PORTS, VAL) .
```

The (equationally defined) function `executeTransition` executes a transition with `PORTS` the states of the ports and `VAL` the values of the state variables. The function returns a triple `transResult`(p, σ, t), where p is the state of the ports after the execution, σ is the resulting values of the state variables, and t is the sum of the `delays` in the transition actions. The transitions are modeled as a multiset; therefore, *any* enabled transition can be applied in the rule.

Tool Support. This first work did not fully integrate formal analysis into the OSATE tool environment for AADL: The *ADDL2Maude* OSATE plugin used OSATE's code generation facilities to automatically generate Real-Time Maude models from AADL specifications. However, the analysis required running commands in Real-Time Maude. Nevertheless, the user did not have to know the Real-Time Maude representation of her AADL model, since we provided convenient syntax for defining state patterns. The term

```
value of v in component fullComponentName in globalComponent
```

gives the value of the state variable v in the thread identified by the full component name *fullComponentName* in the system in state *globalComponent*, and

```
location of component fullComponentName in globalComponent
```

gives the current location/state of the transition system in the given thread. For LTL model checking, we pre-defined parametric atomic propositions such as *full thread name @ location*, which holds when the thread is in state *location*.

Applications. We used AADL2Maude and Real-Time Maude analysis of the generated model to analyze an AADL model developed at UIUC of a network of medical devices consisting of a *controller*, a *ventilator machine* that assists a patient's breathing during surgery, and an *X-ray* device. When a button is pushed to take an X-ray, and the ventilator machine has *not* paused in the past 10 min, the ventilator machine pauses for two seconds, starting one second after the button is pushed, and the X-ray should be taken after two seconds.

To execute the system, we added a *test activator* that pushes the button every second. We then used reachability analysis to check whether an *undesired* state, where the X-ray thread is in state `xray` (X-ray being taken) and the ventilator is *not* in state `paused`, can be reached (which leads to blurry pictures):

```
Maude> (search [1]  initialize({MAIN system Wholesys . impl}) =>* {CONF}
         such that ((location of component (MAIN -> Xray -> xmPr -> xmTh)
                     in CONF) == xray and (location of component (MAIN ->
                     Ventilator -> vmPr -> vmTh) in CONF) =/= paused) .)

Solution 1        CONF --> ...
```

(The unexpected result showed that such a bad state can indeed be reached.) We used time-bounded LTL model checking to verify that an X-ray must be taken within three seconds of the start of the system (surprisingly, this command returned a counterexample revealing a subtle and previously unknown error):

```
Maude> (mc initialize({MAIN system Wholesys . impl}) |=t
            <> ((MAIN -> Xray -> xmPr -> xmTh) @ xray) in time <= 3 .)
```

We also wanted to verify an AADL model of the *active standby* system. However, no serious analysis finished within reasonable time, as explained in [56].

Nevertheless, this work defines a formal semantics of a useful subset of AADL, and could be used to find subtle flaws in existing AADL models.

5   The PALS Synchronizer for CPSs

Model checking even simplified versions (e.g., without clock drifts and communication delays) of the *active standby* system turned out to be unfeasible. Even *designing* the system, to ensure that messages are read in the correct round in the presence of fast and slow local clocks and message delays, required modeling buffering of both incoming and outgoing messages and subtle use of (local-clock-based) timers to read and transmit messages at the correct times.

The active standby system shares many characteristics with (distributed) CPSs in fields such as avionics, cars, robotics, and other control systems:

- The "underlying logic" is *synchronous*: At the beginning/end of each "period" all components should in lockstep read incoming messages and perform local transitions, which change the state of each component and generate messages for the *next* iteration. However, the system has to be realized in a distributed setting, with message delays, imprecise local clocks, execution times, etc.
- Cars, airplanes, factories, robots, etc., typically use dedicated local networks where network delays are bounded.
- Clock synchronization is well understood; e.g., the Precision Time Protocol (IEEE 1588) achieves sub-microsecond clock accuracy on local area networks [29]. We can therefore often give a bound ϵ on the skew of the clocks.

All of this inspired us and colleagues at UIUC and Rockwell Collins to define the PALS ("physically asynchronous, logically synchronous") *formal pattern* [51] to greatly simplify both the design and the verification of these "logically synchronous" (distributed) CPSs when the infrastructure guarantees bounds μ, α, and ϵ on the communication delays, transition computation times, and clock drifts, respectively [3,52,66]. For such bounds $\Gamma = (\mu, \alpha, \epsilon)$, PALS is a mapping

$$(SD, p, \Gamma) \mapsto \text{PALS}(SD, p, \Gamma)$$

where SD is the underlying synchronous design, p is the period of the components, and $\text{PALS}(SD, p, \Gamma)$ is the corresponding distributed system model. SD is formalized as the synchronous composition of a collection of state

("Mealy") machines, each with typed input and output ports, whose ports are connected by a *wiring diagram*. Each machine M has a transition relation $\delta_M \subseteq (inputs \times State) \times (State, outputs)$ and performs one transition in each iteration of the system. The distributed real-time system $PALS(SD, p, \Gamma)$ is formalized in [52] using Real-Time Maude. It is proved in [52] that as long as the *PALS constraint* $p \geq \mu + 2\epsilon + \max(2\epsilon, \alpha)$ is satisfied (the difference between two clocks is always less than 2ϵ), then SD and $PALS(SD, p, \Gamma)$ satisfy the same CTL^* formulas. (The constraint in [52] also takes into account the minimum network delay.)

It is therefore sufficient to specify and verify the much simpler underlying synchronous model. PALS then provides the corresponding distributed model, which satisfies the same properties as the synchronous design.

Synchronizers (like GALS and LTTA) relating synchronous and asynchronous systems are well known. However, very few, such as the *time-triggered architecture* (TTA) [38,65], target CPSs with bounded network delays and clock skews, and therefore do not guarantee timeliness of the resulting asynchronous system (see [52] for details). TTA has similar assumptions as PALS, but has a significantly longer optimal period p than PALS, and some other differences [13,52,68].

Effectiveness and Application. In [52] we modeled the synchronous PALS design of active standby in Maude, as well as a much simplified asynchronous model, with perfect clocks and no execution times: The synchronous model has **185** reachable states and could be model checked in less than a second. The asynchronous model has **3,047,832** reachable states when message delays are 0; a simple reachability command explores these states in 2,000 s. When the message delay could be 0 or 1, attempts at exploring the state space aborted.

With PALS we could easily analyze the requirements of active standby using LTL model checking: most did not hold. Inspecting the counterexamples, we came up with, and verified, modified properties, which were the properties ones found by the Rockwell Collins team using NuSMV and the PALS methodology.

6 Synchronous AADL

Synchronous AADL [14] is a subset of AADL in which synchronous designs in general, as well as synchronous PALS models, can be specified. Synchronous AADL is defined as a behavioral subset of AADL explained in Sect. 4, together with syntactic constraints to identify AADL models that can be considered synchronous: e.g., threads have periodic dispatch; components have only data ports; and connections between threads are delayed. Each AADL construct in the subset has the same meaning in AADL and Synchronous AADL, and properties specific to Synchronous AADL are declared using the property set SynchAADL.

The formal semantics of Synchronous AADL is defined in Real-Time Maude, but now specifies the synchronous composition of AADL components.

Real-Time Maude analysis of Synchronous AADL models is integrated into OSATE using the SynchAADL2Maude plugin [18]. Given a Synchronous AADL model *SD*, the tool checks whether *SD* is a valid Synchronous AADL model, generates the corresponding Real-Time Maude model, and invokes Real-Time Maude to analyze whether *SD* satisfies given LTL properties. SynchAADL2Maude provides predefined atomic propositions to easily specify system properties, and LTL properties to be analyzed are managed by an XML file. Figure 5 shows an example of such properties and the SynchAADL2Maude window for the active standby system, where the analysis results are shown in the Maude Console.

7   Multirate PALS and MR-SynchAADL

Whereas PALS, TTA, and synchronous systems in general require that all components have the same period, different ("off-the-shelf") controllers often operate at different frequencies, yet need to synchronize, as in the airplane turning system in Sect. 3. We have therefore extended PALS, the Synchronous AADL modeling language, and the SynchAADL2Maude tool to the multirate setting [11,12,17].

Composing components with different periods is tricky, since a controller with period 30ms receives and sends messages every 30ms, whereas one with period with 50ms only does so every 50ms. We address this as follows:

- A component may only communicate with components whose period is a multiple of its own period (possibly through a hierarchy); or vice versa.
- User-defined *input adaptors* turn one output value from a slow component to *k* input values for a connecting component which is *k* times faster. Likewise, the slow component's input adaptor turns *k* outputs from the fast component into a single input to the slow component.

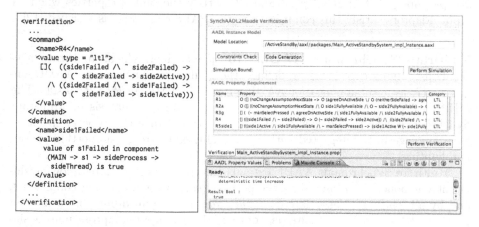

Fig. 5. Properties in the XML format (left) and SynchAADL2Maude window (right).

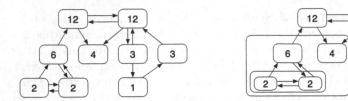

Fig. 6. A hierarchical multirate system.   **Fig. 7.** A multirate synchronous design.

Since a synchronous system can be represented as a single machine [12,52], we can have hierarchical models. Figure 6 shows an example of multirate systems, and Fig. 7 shows its corresponding hierarchical synchronous design, where each machine and its local environment are annotated with its period.

Multirate PALS. We "slow down" the faster components, so that all components in a multirate synchronous design operate in lock-step as in the single-rate case. A fast component slowed down by a factor k performs k *internal transitions* during one (slow) period. It consumes k inputs and produces k outputs at each port in each slow step. A slow controller, which should only read *one* input (in each input port) during such a slow step, therefore uses an *input adaptor* to transform such a k-tuple output into a single input value; and vice versa for slow-to-fast connections.

For a multirate synchronous design SD, bounds Γ, and a global period p, Multirate PALS gives the distributed real-time system $\mathcal{MPALS}(SD, p, \Gamma)$ where each component operates according to its own period [11,12]. However, a fast machine may not be able to finish all of its k internal transitions in a slow period p before the output messages must be sent. If only $k_f < k$ outputs can be sent before the deadline, the input adaptor must ignore the last $k - k_f$ values in a k-tuple input. $\mathcal{MPALS}(SD, p, \Gamma)$ and SD satisfy the same properties when all input adaptors satisfy this condition and the PALS constraint is satisfied [12].

Multirate Synchronous AADL. To specify hierarchical multirate synchronous designs in AADL, we defined the Multirate Synchronous AADL language [17] as a sublanguage of AADL with a property set MR_SynchAADL. The MR-SynchAADL plugin supports modeling and formal analysis of Multirate Synchronous AADL models within OSATE, with a dedicated property specification language fully integrated with the OSATE editor (e.g., supporting syntax highlighting).

Multirate Synchronous AADL extends Synchronous AADL by allowing different components to have different periods (satisfying the above conditions), and by allowing us to associate an *input adaptor* α to an input port using the property MR_SynchAADL::InputAdaptor=> α. 1-to-k input adaptors map a single value to a k-vector of values, and k-to-1 input adaptors map a k-vector of values to a single value. Multirate Synchronous AADL provides a collection of predefined input adaptors, including repeat_input—mapping v to $(v, v, \ldots, v)$—, and last—mapping $(v_1, \ldots, v_k)$ to v_k—, and so on. The formal semantics of Multirate Synchronous AADL is also defined in Real-TIme Maude.

Application. In [9,12,17], we modeled and analyzed the Multirate PALS synchronous model of the airplane example in Sect. 3 using Real-Time Maude. The continuous behavior was numerically approximated using Euler's method. The formal analysis revealed a flaw that caused an unsafe turn. This led to a redesign of the system, which was verified using model checking. Again, (bounded) model checking of a highly simplified *asynchronous* model was unfeasible due to the many interleavings caused. The airplane example was also modeled in Multirate Synchronous AADL and analyzed using MR-SynchAADL in [17].

8   Hybrid PALS and HybridSynchAADL

CPS controllers may interact with *physical* environments, whose *continuous* dynamics can be modeled as ordinary differential equations (ODEs). To precisely analyze such CPSs with the PALS methodology, we have developed the Hybrid PALS synchronizer [16] and the HybridSynchAADL language and tool [40,41].

Hybrid PALS extends Multirate PALS to CPSs with continuous dynamics. In contrast to (Multirate) PALS, we cannot abstract away the times at which physical states are sampled and actuated. We must also to take into account *imprecise local clocks*, since such *sampling and actuating times* depend on them.

A formal analysis of such CPSs involves an infinite number of continuous trajectories depending on imprecise local clocks. We therefore use Maude combined with SMT solving to *symbolically* encode all possible continuous behaviors, and define a symbolic (Maude with SMT) semantics for HybridSynchAADL language.

Hybrid PALS. We consider multirate systems consisting of discrete controllers and physical environments, as shown in Fig. 8. A controller M is a nondeterministic machine parameterized by any behavior of its physical environment E_M. Environments can also be physically correlated (dashed lines in Fig. 8), meaning that changes in one environment immediately affect another.

A state of a physical environment E_M is a valuation of real-valued parameters $\vec{x} = (x_1, \ldots, x_l)$. The continuous behavior of E_M is modeled by systems of ODEs that specify different trajectories of $\vec{x}$ over time. A control command from its controller M defines *which* trajectories E_M follows. In Fig. 9, E_M follows trajectory τ_1 from state v_1 for duration $t_2 - t_1$ when command a_1 is received.

Let $c_M(i)$ denote the *global time* when the i-th period of a controller M begins according to its local clock. M samples the state of E_M at time $c_M(i) + t_s$, where t_s is a value in its sampling time interval. M then performs a transition and determines a new control command. E_M receives the new command at time $c_M(i) + t_a$, where t_a is a value in its actuating (or response) time interval.

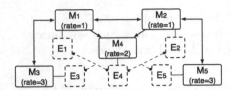

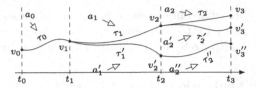

Fig. 8. A hybrid multirate system.   **Fig. 9.** A controlled physical environment.

A hybrid synchronous design $SD \upharpoonright E$ is composed of a multirate synchronous design SD and a collection E of physical environments. The "discrete" behavior is given by the synchronous design SD *restricted by* the behavior of E, and the "continuous" behavior is given by a set of trajectories of E *realizable by* control commands from SD.

For a hybrid synchronous design $SD \upharpoonright E$, a global period p, and bounds Γ, Hybrid PALS produces the distributed *hybrid* system $\mathcal{M}\text{PALS}(SD, p, \Gamma) \upharpoonright E$, where each controller interacts with its physical environment in E. Hybrid PALS ensures that both the synchronous design $SD \upharpoonright E$ and the distributed hybrid system $\mathcal{M}\text{PALS}(SD, p, \Gamma) \upharpoonright E$ exhibit exactly the same set of trajectories of E.

HybridSynchAADL. The HybridSynchAADL language and tool [39–41] extends (Multirate) Synchronous AADL and MR-SynchAADL to hybrid synchronous designs. A physical environment is modeled as an environment component in HybridSynchAADL, and can have different *modes* to specify different trajectories. The continuous dynamics in is specified using the property `Hybrid_SynchAADL::ContinuousDynamics`. Figure 10 shows an example of an environment components with two modes. A controller is an ordinary software component. It declares the three properties `Hybrid_SynchAADL::Max_Clock_Deviation`, `Hybrid_SynchAADL::Sampling_Time`, and `Hybrid_SynchAADL::Response_Time` to specify the maximal clock skew, and sampling and actuating time intervals.

```
system RoomEnv
  features
    temp: out data port Base_Types::Float;
    power: in data port Base_Types::Float;
    on_ctrl: in event port;
    off_ctrl: in event port;
  properties
    Hybrid_SynchAADL::isEnvironment => true;
end RoomEnv;
```

```
system implementation RoomEnv.impl
  subcomponents x: data Base_Types::Float; p: data Base_Types::Float;
  connections   C: port x -> temp;      R: port power -> p;
  modes         hOff: initial mode;
                hOff -[on_ctrl]-> hOn;    hOn -[off_ctrl]-> hOff;
  properties    Hybrid_SynchAADL::ContinuousDynamics =>
                "x(t)= x(0)- 0.1*(x(0)- p/0.1)* t;"  in modes (hOn),
                "x(t)= x(0)*(1 - 0.1 * t);"  in modes (hOff);
end RoomEnv.impl;
```

Fig. 10. An environment component.

We defined the Maude+SMT semantics of HybridSynchAADL for *single-rate* designs. A *constrained object* of the form $\phi \parallel obj$ symbolically represents (infinitely many) instances of *obj* satisfying the constraint ϕ. The behavior of individual components for one synchronous iteration is specified by the operation `executeStep`, defined by rewrite rules on constrained terms. The synchronous step of the entire system is then formalized by the following rule, which captures all possible behaviors from any instance of $\{\phi \mathrel{||} obj\}$:

```
crl [step]: {PHI || < C : System | features : none >} => {PHI' || OBJ'}
  if executeStep(PHI || < C : System | >) => PHI' || OBJ' .
```

Tool Support. HybridSynchAADL supports Maude+SMT reachability analysis of single-rate hybrid synchronous designs within OSATE. The property specification language allows the user to specify reachability properties of the form

$$\text{reachability } \varphi_{init} \texttt{ ==> } \varphi_{goal} \text{ in time } \tau,$$

which holds if a state satisfying φ_{goal} is reachable from a state satisfying φ_{init} within time τ. The tool provides three analysis methods: symbolic reachability analysis using Maude+SMT; randomized simulation, which repeatedly runs the model by randomly choosing concrete data values, sampling and actuating times, etc.; and portfolio analysis, which invokes randomized simulation and symbolic reachability analysis in parallel using multithreading.

Fig. 11. HybridSynchAADL window in OSATE.

Applications. We used HybridSynchAADL to model and analyze distributed drones that collaborate to perform rendezvous, formation control, and packet delivery [39–41]. For rendezvous, we symbolically analyzed two time-bounded properties: drones do not collide within time 500 (`[safety]`), and all drones gather together within time 500 (`[rendezvous]`). These properties use three user-defined propositions `init`, `collide`, and `gather`, and are defined as follows in our property specification language:

```
invariant [safety]: ?initial ==> not ?collide in time 500;
reachability [rendezvous]: ?initial ==> ?gather in time 500;
```

Figure 11 shows the tool interface that is fully integrated into OSATE, where the analysis results are shown in the `Result` view. We compare the performance of HybridSynchAADL's symbolic analysis with four hybrid systems reachability analysis tools, such as [25,26,33,37]. The experiments showed that (in most cases) HybridSynchAADL outperforms these state-of-the-art tools.

9   Future Research Directions

We briefly mention some future research directions on three key components of our formal model engineering approach to developing reliable distributed CPSs: (1) *synchronizers* to reduce the design and analysis complexity; (2) "AADL-based" *modeling languages* with formal semantics; and (3) *formal analysis techniques*.

Synchronizers. The MSYNC synchronizer [13] generalizes Multirate PALS and TTA, and should be extended to CPSs with continuous dynamics. We should also extend other synchronizers, such as loosely time-triggered architectures [23, 70], to the multirate setting with continuous dynamics.

AADL-based Modeling Languages with Formal Semantics. We should extend the Maude-SMT semantics of Hybrid Synchronous AADL to multirate systems. The Maude-SMT analysis in HybridSynchAADL can only deal with polynomial continuous dynamics. To support general classes of ODEs, we should integrate Maude with ODE solvers, such as dReal [34]. The airplane turning algorithm in Sect. 3 could then be symbolically analyzed using HybridSynchAADL. We should also support more AADL language constructs and the modeling of (Hybrid) MSYNC models. Language extensions should be guided by more applications, such as the steam-boiler controller [1] and aerospace systems [24,47].

Formal Analysis. We should support formal analysis methods beyond explicit-state LTL/TCTL model checking and symbolic reachability analysis, e.g., STL model checking [10,42,73] and statistical model checking [4,45,64].

HybridSynchAADL employs state merging [19,20] to improve the performance of Maude+SMT symbolic analysis. We can further improve the performance by applying incremental rewriting modulo SMT [71]. Abstracting away imprecise local clocks in Hybrid PALS, e.g., by over-approximation, should also significantly improve the scalability of the analysis.

Maude+SMT analysis opens up the possibility of having *parametric* initial states, and automatically *synthesizing* parameter values that make the system satisfy desired properties, as we have done for timed automata and Petri nets [5, 6]. We should explore such parameter synthesis for logically synchronous CPSs.

10   Concluding Remarks

Equipped with the expressive Maude and Real-Time Maude formalisms—which should be able to capture the semantics of industrial modeling languages—whose tools nevertheless provide powerful *automatic* formal analyses, our goal is to integrate formal analysis of CPSs into modeling tools used in industry, so that a designer can formally analyze her models without knowing formal methods.

Motivated by projects at UIUC, and having access to AADL models of a network of medical devices and of a Rockwell Collins-developed avionics system, we decided to target the component-based industrial modeling standard AADL. In this paper we give, for the first time, an overview of this effort.

We first gave a Real-Time Maude semantics to a significant subset of the "software" parts of AADL, and used Real-Time Maude analysis to find subtle flaws in the AADL model of the medical system. However, we could not analyze the avionics system, whose NuSMV model checking also caused major problems.

Observing that the active standby system and many other CPSs have "logically synchronous" designs but have to be realized as distributed systems on local area networks, we developed the PALS *formal pattern* as one of the first "synchronizers for CPSs."[1] It is therefore sufficient to model and verify the much simpler underlying synchronous designs. In particular, PALS reduced the intractable task of model checking active standby to one that could be done in less than a second.

To make PALS-based formal analysis of "logically synchronous" distributed systems on local area networks available to AADL modelers, we identified an annotated sublanguage of AADL, called Synchronous AADL, that can be used by AADL modelers not only to model synchronous PALS models, but also synchronous systems in general, and integrated Real-Time Maude analysis of Synchronous AADL models into the OSATE tool environment for AADL.

Since controllers may operate at different frequencies, yet need to synchronize, we extended PALS and Synchronous AADL to the multirate setting. Multirate PALS, introduced in a FACS 2012 paper, was, to the best of our knowledge, the first synchronizer for multirate CPSs. We applied our methodology to an airplane turning control algorithm and found that it did not provide a safe yaw.

The airplane turning example emphasized that many CPSs interact with *continuous* environments. We therefore extended our methods to hybrid systems. Hybrid PALS cannot abstract from the times when sensing and actuating such environments happen, which depend on local clocks with bounded but unknown skews. The recent integration of SMT solving with Maude allows us to capture all the possible continuous behaviors symbolically. We defined the HybridSynchAADL language and integrated Maude+SMT-based simulation and reachability analysis into OSATE. Experiments on collaborating UAVs showed that HybridSynchAADL analysis in many cases outperforms state-of-the-art hybrid systems reachability tools such as HyComp, Flow*, SpaceEx, and dReach [41].

Finally, we outlined some future directions in this line of work.

Acknowledgments. We thank Olga Kouchnarenko and the organizers of FACS 2023 for inviting us to present this work summarizing some of our "software component-based" research to celebrate the 20th anniversary of the FACS conference. In this paper

---

[1] We do not have space to discuss related work in this overview paper, but our papers have extensive discussions of related work that justify our claims.

we report on research initiated by some of us and Darren Cofer and Steven Miller at Rockwell Collins and José Meseguer and Lui Sha at University of Illinois at Urbana-Champaign; we sincerely thank them all, as well as the coauthors of all the papers summarized in this paper. Bae was supported by the National Research Foundation of Korea(NRF) grants funded by the Korea government(MSIT) (No. 2021R1A5A1021944 and No. RS-2023-00251577).

References

1. Abrial, J.-R., Börger, E., Langmaack, H. (eds.): Formal Methods for Industrial Applications. LNCS, vol. 1165. Springer, Heidelberg (1996). https://doi.org/10.1007/BFb0027227
2. Al-Nayeem, A., Sha, L., Cofer, D.D., Miller, S.M.: Pattern-based composition and analysis of virtually synchronized real-time distributed systems. In: 2012 IEEE/ACM Third International Conference on Cyber-Physical Systems, pp. 65–74. IEEE (2012)
3. Al-Nayeem, A., Sun, M., Qiu, X., Sha, L., Miller, S.P., Cofer, D.D.: A formal architecture pattern for real-time distributed systems. In: Proceedings of RTSS, pp. 161–170. IEEE, USA (2009)
4. AlTurki, M., Meseguer, J.: PVESTA: a parallel statistical model checking and quantitative analysis tool. In: Corradini, A., Klin, B., Cîrstea, C. (eds.) CALCO 2011. LNCS, vol. 6859, pp. 386–392. Springer, Heidelberg (2011). https://doi.org/10.1007/978-3-642-22944-2_28
5. Arias, J., Bae, K., Olarte, C., Ölveczky, P.C., Petrucci, L., Rømming, F.: Rewriting logic semantics and symbolic analysis for parametric timed automata. In: 8th ACM SIGPLAN International Workshop on Formal Techniques for Safety-Critical Systems (FTSCS 2022), pp. 3–15. ACM (2022)
6. Arias, J., Bae, K., Olarte, C., Ölveczky, P.C., Petrucci, L., Rømming, F.: Symbolic analysis and parameter synthesis for time Petri nets using Maude and SMT solving. In: Gomes, L., Lorenz, R. (eds.) Application and Theory of Petri Nets and Concurrency. PETRI NETS 2023. LNCS, vol. 13929. Springer, Cham (2023). https://doi.org/10.1007/978-3-031-33620-1_20
7. Arney, D., Jetley, R., Jones, P., Lee, I., Sokolsky, O.: Formal methods based development of a PCA infusion pump reference model: Generic infusion pump (GIP) project. In: HCMDSS-MDPnP, pp. 23–33. IEEE (2007)
8. Bae, K., Ölveczky, P.C., Feng, T.H., Tripakis, S.: Verifying Ptolemy II discrete-event models using Real-Time Maude. In: Breitman, K., Cavalcanti, A. (eds.) ICFEM 2009. LNCS, vol. 5885, pp. 717–736. Springer, Heidelberg (2009). https://doi.org/10.1007/978-3-642-10373-5_37
9. Bae, K., Krisiloff, J., Meseguer, J., Ölveczky, P.C.: Designing and verifying distributed cyber-physical systems using Multirate PALS: an airplane turning control system case study. Sci. Comput. Program. 103, 13–50 (2015)
10. Bae, K., Lee, J.: Bounded model checking of signal temporal logic properties using syntactic separation. Proc. ACM Program. Lang. 3(POPL), 1–30 (2019)
11. Bae, K., Meseguer, J., Ölveczky, P.C.: Formal patterns for multi-rate distributed real-time systems. In: Păsăreanu, C.S., Salaün, G. (eds.) FACS 2012. LNCS, vol. 7684, pp. 1–18. Springer, Heidelberg (2013). https://doi.org/10.1007/978-3-642-35861-6_1
12. Bae, K., Meseguer, J., Ölveczky, P.C.: Formal patterns for multirate distributed real-time systems. Sci. Comput. Program. 91, 3–44 (2014)

13. Bae, K., Ölveczky, P.C.: MSYNC: a generalized formal design pattern for virtually synchronous multirate cyber-physical systems. ACM Trans. Embed. Comput. Syst. (TECS) **20**(5s), 1–26 (2021)

14. Bae, K., Ölveczky, P.C., Al-Nayeem, A., Meseguer, J.: Synchronous AADL and its formal analysis in Real-Time Maude. In: Qin, S., Qiu, Z. (eds.) ICFEM 2011. LNCS, vol. 6991, pp. 651–667. Springer, Heidelberg (2011). https://doi.org/10.1007/978-3-642-24559-6_43

15. Bae, K., Ölveczky, P.C., Feng, T.H., Lee, E.A., Tripakis, S.: Verifying hierarchical Ptolemy II discrete-event models using Real-Time Maude. Sci. Comput. Program. **77**(12), 1235–1271 (2012)

16. Bae, K., Ölveczky, P.C., Kong, S., Gao, S., Clarke, E.M.: SMT-based analysis of virtually synchronous distributed hybrid systems. In: Proceedings of HSCC, pp. 145–154. ACM, New York, NY, USA (2016)

17. Bae, K., Ölveczky, P.C., Meseguer, J.: Definition, semantics, and analysis of Multirate Synchronous AADL. In: Jones, C., Pihlajasaari, P., Sun, J. (eds.) FM 2014. LNCS, vol. 8442, pp. 94–109. Springer, Cham (2014). https://doi.org/10.1007/978-3-319-06410-9_7

18. Bae, K., Ölveczky, P.C., Meseguer, J., Al-Nayeem, A.: The SynchAADL2Maude tool. In: de Lara, J., Zisman, A. (eds.) FASE 2012. LNCS, vol. 7212, pp. 59–62. Springer, Heidelberg (2012). https://doi.org/10.1007/978-3-642-28872-2_4

19. Bae, K., Rocha, C.: Guarded terms for rewriting modulo SMT. In: Proença, J., Lumpe, M. (eds.) FACS 2017. LNCS, vol. 10487, pp. 78–97. Springer, Cham (2017). https://doi.org/10.1007/978-3-319-68034-7_5

20. Bae, K., Rocha, C.: Symbolic state space reduction with guarded terms for rewriting modulo SMT. Sci. Comput. Program. **178**, 20–42 (2019)

21. Barrett, C., et al.: CVC4. In: Gopalakrishnan, G., Qadeer, S. (eds.) CAV 2011. LNCS, vol. 6806, pp. 171–177. Springer, Heidelberg (2011). https://doi.org/10.1007/978-3-642-22110-1_14

22. Barrett, C., Stump, A., Tinelli, C., et al.: The SMT-LIB standard: Version 2.0. In: SMT. vol. 13, p. 14 (2010)

23. Baudart, G., Benveniste, A., Bourke, T.: Loosely time-triggered architectures: improvements and comparisons. ACM Trans. Embed. Comput. Syst. (TECS) **15**(4), 1–26 (2016)

24. Bozzano, M., Bruintjes, H., Cimatti, A., Katoen, J.-P., Noll, T., Tonetta, S.: Formal methods for aerospace systems. In: Nakajima, S., Talpin, J.-P., Toyoshima, M., Yu, H. (eds.) Cyber-Physical System Design from an Architecture Analysis Viewpoint, pp. 133–159. Springer, Singapore (2017). https://doi.org/10.1007/978-981-10-4436-6_6

25. Chen, X., Ábrahám, E., Sankaranarayanan, S.: Flow*: an analyzer for non-linear hybrid systems. In: Sharygina, N., Veith, H. (eds.) CAV 2013. LNCS, vol. 8044, pp. 258–263. Springer, Heidelberg (2013). https://doi.org/10.1007/978-3-642-39799-8_18

26. Cimatti, A., Griggio, A., Mover, S., Tonetta, S.: HyComp: an SMT-based model checker for hybrid systems. In: Baier, C., Tinelli, C. (eds.) TACAS 2015. LNCS, vol. 9035, pp. 52–67. Springer, Heidelberg (2015). https://doi.org/10.1007/978-3-662-46681-0_4

27. Clavel, M., et al.: All About Maude - A High-Performance Logical Framework. LNCS, vol. 4350. Springer, Heidelberg (2007). https://doi.org/10.1007/978-3-540-71999-1

28. Dutertre, B.: Yices 2.2. In: Biere, A., Bloem, R. (eds.) CAV 2014. LNCS, vol. 8559, pp. 737–744. Springer, Cham (2014). https://doi.org/10.1007/978-3-319-08867-9_49

29. Eidson, J.: https://www.nist.gov/document/tutorial-basicpdf/ (2005). Accessed 16 Jul 2023

30. Eidson, J.C., Fischer, M., White, J.: IEEE-1588™ standard for a precision clock synchronization protocol for networked measurement and control systems. In: Proceedings of the 34th Annual Precise Time and Time Interval Systems and Applications Meeting, pp. 243–254 (2002)

31. Feiler, P.H., Gluch, D.P.: Model-Based Engineering with AADL: An Introduction to the SAE Architecture Analysis and Design Language. Addison-Wesley, USA (2012)

32. França, R., Bodeveix, J.P., Filali, M., Rolland, J.F., Chemouil, D., Thomas, D.: The AADL Behaviour Annex - experiments and roadmap. In: Proceedings of ICECCS 2007. IEEE, USA (2007)

33. Frehse, G., et al.: SpaceEx: scalable verification of hybrid systems. In: Gopalakrishnan, G., Qadeer, S. (eds.) CAV 2011. LNCS, vol. 6806, pp. 379–395. Springer, Heidelberg (2011). https://doi.org/10.1007/978-3-642-22110-1_30

34. Gao, S., Kong, S., Clarke, E.M.: dReal: an SMT solver for nonlinear theories over the reals. In: Bonacina, M.P. (ed.) CADE 2013. LNCS (LNAI), vol. 7898, pp. 208–214. Springer, Heidelberg (2013). https://doi.org/10.1007/978-3-642-38574-2_14

35. Goguen, J., Meseguer, J.: Order-sorted algebra I: equational deduction for multiple inheritance, overloading, exceptions and partial operations. Theoret. Comput. Sci. **105**, 217–273 (1992)

36. Kim, C., Sun, M., Mohan, S., Yun, H., Sha, L., Abdelzaher, T.F.: A framework for the safe interoperability of medical devices in the presence of network failures. In: ICCPS, pp. 149–158 (2010)

37. Kong, S., Gao, S., Chen, W., Clarke, E.: dReach: δ-reachability analysis for hybrid systems. In: Baier, C., Tinelli, C. (eds.) TACAS 2015. LNCS, vol. 9035, pp. 200–205. Springer, Heidelberg (2015). https://doi.org/10.1007/978-3-662-46681-0_15

38. Kopetz, H., Bauer, G.: The time-triggered architecture. Proc. IEEE **91**(1), 112–126 (2003)

39. Lee, J., Bae, K., Ölveczky, P.C.: An extension of HybridSynchAADL and its application to collaborating autonomous UAVs. In: Margaria, T., Steffen, B. (eds.) Leveraging Applications of Formal Methods, Verification and Validation. Adaptation and Learning. ISoLA 2022. LNCS, vol. 13703. Springer, Cham (2022). https://doi.org/10.1007/978-3-031-19759-8_4

40. Lee, J., Bae, K., Ölveczky, P.C., Kim, S., Kang, M.: Modeling and formal analysis of virtually synchronous cyber-physical systems in AADL. Int. J. Softw. Tools Technol. Transfer **24**(6), 911–948 (2022)

41. Lee, J., Kim, S., Bae, K., Ölveczky, P.C.: HYBRID SYNCHAADL: modeling and formal analysis of virtually synchronous CPSs in AADL. In: Silva, A., Leino, K.R.M. (eds.) CAV 2021. LNCS, vol. 12759, pp. 491–504. Springer, Cham (2021). https://doi.org/10.1007/978-3-030-81685-8_23

42. Lee, J., Yu, G., Bae, K.: Efficient SMT-based model checking for signal temporal logic. In: 2021 36th IEEE/ACM International Conference on Automated Software Engineering (ASE), pp. 343–354. IEEE (2021)

43. Leen, G., Heffernan, D., Dunne, A.: Digital networks in the automotive vehicle. Comput. Control Eng. J. **10**(6), 257–266 (1999)

44. Lepri, D., Ábrahám, E., Ölveczky, P.C.: Sound and complete timed CTL model checking of timed Kripke structures and real-time rewrite theories. Sci. Comput. Program. **99**, 128–192 (2015)

45. Liu, S., Meseguer, J., Ölveczky, P.C., Zhang, M., Basin, D.: Bridging the semantic gap between qualitative and quantitative models of distributed systems. Proc. ACM Program. Lang. **6**(OOPSLA2), 315–344 (2022)

46. Liu, S., Ölveczky, P.C., Meseguer, J.: Modeling and analyzing mobile ad hoc networks in Real-Time Maude. J. Log. Algebraic Methods Program. **85**(1), 34–66 (2016). https://doi.org/10.1016/j.jlamp.2015.05.002

47. Mavridou, A., Stachtiari, E., Bliudze, S., Ivanov, A., Katsaros, P., Sifakis, J.: Architecture-based design: a satellite on-board software case study. In: Kouchnarenko, O., Khosravi, R. (eds.) FACS 2016. LNCS, vol. 10231, pp. 260–279. Springer, Cham (2017). https://doi.org/10.1007/978-3-319-57666-4_16

48. Meseguer, J.: Conditional rewriting logic as a unified model of concurrency. Theoret. Comput. Sci. **96**(1), 73–155 (1992)

49. Meseguer, J.: Taming distributed system complexity through formal patterns. In: Arbab, F., Ölveczky, P.C. (eds.) FACS 2011. LNCS, vol. 7253, pp. 1–2. Springer, Heidelberg (2012). https://doi.org/10.1007/978-3-642-35743-5_1

50. Meseguer, J.: Twenty years of rewriting logic. J. Logic Algebraic Program. **81**(7), 721–781 (2012)

51. Meseguer, J.: Taming distributed system complexity through formal patterns. Sci. Comput. Program. **83**, 3–34 (2014)

52. Meseguer, J., Ölveczky, P.C.: Formalization and correctness of the PALS architectural pattern for distributed real-time systems. Theoret. Comput. Sci. **451**, 1–37 (2012)

53. Meseguer, J., Roşu, G.: The rewriting logic semantics project: a progress report. Inf. Comput. **231**, 38–69 (2013)

54. Miller, O., Oofer, D., Oha, L., Meseguer, J., Al-Nayeem, A.: Implementing logical synchrony in integrated modular avionics. In: Proceedings of IEEE/AIAA 28th Digital Avionics Systems Conference. IEEE, USA (2009)

55. Ölveczky, P.C., Boronat, A., Meseguer, J.: Formal semantics and analysis of behavioral AADL models in Real-Time Maude. In: Hatcliff, J., Zucca, E. (eds.) FMOODS/FORTE -2010. LNCS, vol. 6117, pp. 47–62. Springer, Heidelberg (2010). https://doi.org/10.1007/978-3-642-13464-7_5

56. Ölveczky, P.C., Boronat, A., Meseguer, J., Pek, E.: Formal semantics and analysis of behavioral AADL models in Real-Time Maude. https://olveczky.se/RealTimeMaude/AADL/webTechRep.pdf (2010)

57. Ölveczky, P.C., Meseguer, J.: Semantics and pragmatics of Real-Time Maude. High.-Order Symbolic Compu. **20**(1–2), 161–196 (2007)

58. Ölveczky, P.C., Meseguer, J.: The Real-Time Maude Tool. In: Ramakrishnan, C.R., Rehof, J. (eds.) TACAS 2008. LNCS, vol. 4963, pp. 332–336. Springer, Heidelberg (2008). https://doi.org/10.1007/978-3-540-78800-3_23

59. Ölveczky, P.C.: Real-Time Maude and its applications. In: Escobar, S. (ed.) WRLA 2014. LNCS, vol. 8663, pp. 42–79. Springer, Cham (2014). https://doi.org/10.1007/978-3-319-12904-4_3

60. Ölveczky, P.C., Thorvaldsen, S.: Formal modeling, performance estimation, and model checking of wireless sensor network algorithms in Real-Time Maude. Theor. Comput. Sci. **410**(2–3), 254–280 (2009). https://doi.org/10.1016/j.tcs.2008.09.022

61. Platzer, A.: Differential dynamic logic for hybrid systems. J. Autom. Reason. **41**(2), 143–189 (2008)

62. Ptolemaeus, C. (ed.): System Design, Modeling, and Simulation using Ptolemy II. Ptolemy.org (2014). http://ptolemy.org/books/Systems
63. Rocha, C., Meseguer, J., Muñoz, C.: Rewriting modulo SMT and open system analysis. J. Logical Algebraic Methods Program. **86**(1), 269–297 (2017)
64. Rubio, R., Martí-Oliet, N., Pita, I., Verdejo, A.: QMaude: quantitative specification and verification in rewriting logic. In: Chechik, M., Katoen, JP., Leucker, M. (eds.) Formal Methods. FM 2023. LNCS, vol. 14000. Springer, Cham (2023). https://doi.org/10.1007/978-3-031-27481-7_15
65. Rushby, J.: Systematic formal verification for fault-tolerant time-triggered algorithms. IEEE Trans. Software Eng. **25**(5), 651–660 (1999)
66. Sha, L., Al-Nayeem, A., Sun, M., Meseguer, J., Ölveczky, P.C.: PALS: Physically asynchronous logically synchronous systems. Tech. rep., Department of Computer Science, University of Illinois at Urbana-Champaign (2009). http://hdl.handle.net/2142/11897
67. Steiner, W., Bauer, G., Hall, B., Paulitsch, M., Varadarajan, S.: TTEthernet dataflow concept. In: 2009 Eighth IEEE International Symposium on Network Computing and Applications, pp. 319–322. IEEE (2009)
68. Steiner, W., Rushby, J.: TTA and PALS: Formally verified design patterns for distributed cyber-physical systems. In: 2011 IEEE/AIAA 30th Digital Avionics Systems Conference, pp. 7B5–1. IEEE (2011)
69. Stevens, B.L., Lewis, F.L.: Aircraft Control and Simulation. John Wiley & Sons (2003)
70. Tripakis, S., Pinello, C., Benveniste, A., Sangiovanni-Vincent, A., Caspi, P., Di Natale, M.: Implementing synchronous models on loosely time triggered architectures. IEEE Trans. Comput. **57**(10), 1300–1314 (2008)
71. Whitters, G., Nigam, V., Talcott, C.L.: Incremental rewriting modulo SMT. In: Pientka, B., Tinelli, C. (eds.) Automated Deduction – CADE 29. CADE 2023. LNCS, vol. 14132. Springer, Cham (2023). https://doi.org/10.1007/978-3-031-38499-8_32
72. Yu, G., Bae, K.: Maude-SE: A tight integration of Maude and SMT solvers. Proc, International Workshop on Rewriting Logic and its Applications (2020)
73. Yu, G., Lee, J., Bae, K.: STLMC: robust STL model checking of hybrid systems using SMT. In: Shoham, S., Vizel, Y. (eds.) Computer Aided Verification. CAV 2022. LNCS, vol. 13371. Springer, Cham (2022). https://doi.org/10.1007/978-3-031-13185-1_26

Challenges Engaging Formal CBSE in Industrial Applications

Yi Li[1] and Meng Sun[2(✉)]

[1] Huawei, Beijing, China
[2] School of Mathematical Sciences, Peking University, Beijing, China
sunm@pku.edu.cn

Abstract. Component-based software engineering (CBSE) is a widely used software development paradigm. With software systems becoming increasingly sophisticated, CBSE provides an effective approach to construct reusable, extensible, and maintainable software systems. Formal verification provides a rigorous and systematic approach to validate the correctness of software systems by mathematically proving properties or checking them exhaustively against specified requirements. Using formal verification techniques in component-based development can further enhance the correctness of the development process. However, the adoption of component-based development supported by formal methods is hardly widespread in the industry. It serves to a limited extent in domains with stringent requirements for safety and reliability. In this paper, we aim to analyze the successful application scenarios of formal methods in component-based development, identify the challenges faced during their application, and explore methods to further broaden their adoption.

Keywords: Formal Methods · Component-based Software Engineering

1 Introduction

The ever-increasing demand for more sophisticated and efficient software systems has necessitated the exploration of novel development methodologies. Among these methodologies, Component-Based Software Engineering (CBSE) [23,39] has emerged as a promising approach that focuses on the development of software systems by assembling pre-existing, self-contained software components. This paradigm shifting from traditional monolithic software development to a component-based approach offers numerous advantages in terms of flexibility, reusability, and modularity.

CBSE does address the issue of code reusability to some extent, and the repeated use and validation of the same components across different projects does improve their reliability. However, it does not fundamentally solve the problem of the correctness of the components and the systems themselves. In recent years, complex software has become increasingly involved in and deeply integrated into people's daily lives, such as autonomous vehicles [13], smart cities [41], and

J. Cámara and S.-S. Jongmans (Eds.): FACS 2023, LNCS 14485, pp. 153–167, 2024.
https://doi.org/10.1007/978-3-031-52183-6_8

smart homes [17]. The correctness of such systems has a significant impact on the safety of human life and property. Formal methods provide a rigorous and systematic approach to software development, ensuring correctness, reliability, and robustness of the system. Therefore, introducing formal methods as an aid in CBSE is of great importance in enhancing software quality and correctness.

A list of impressive research on formal methods in CBSE has been proposed [3,5,23,27,37]. However, in industrial practice, we find that such application is relatively niche. And its usage scenarios are also somewhat limited. In this paper we present the main obstacles, as observed by the authors, that hinder the application of formal methods in the field of component-based software development.

The paper is organized as follows. Section 2 briefly introduces some necessary background knowledge. Section 3 shows some popular industrial examples where formal methods are engaged in CBSE. Section 4 lists major challenges when scaling formal methods to more scenarios. And Sect. 5 discusses and proposes some ideas to tackle the challenges.

2   Background

In this section we introduce some background knowledge to provide readers with a better understanding of the challenges that will be discussed in the following sections.

2.1   Software Development Process

Neither component-based software development nor formal verification is a silver bullet. They are not powerful enough to take over the software development processes [43] widely used in the field today. In typical industrial applications, CBSE is only applied to part of the steps in the whole software development process.

Therefore, for researchers who are interested in applying formal verification techniques or component-based software development approaches in practical applications, it is crucial to understand the actual software development process. Only then can we determine which part of the process our designed methods and developed tools can be applied to and whether they might have any negative impact on the remaining parts of the software development process.

Popular software development processes include the V-model, waterfall model, spiral model, agile development model, and others. Different models are usually associated with different business scenarios. In actual development processes, development teams often customize these processes to align with their specific needs. Therefore, instead of focusing on a specific development model, here we only focus on some common steps in these development models, as shown in Fig. 1.

In the software development process, the steps are all linked with one another. As a result, when we want to use a tool or a methodology to "optimize" some steps, it is extremely important to evaluate its side-effect.

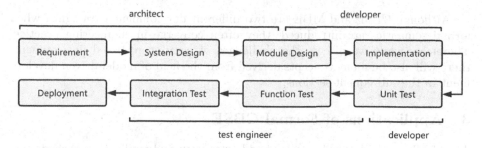

Fig. 1. Software Development Process

For example, if we use a code generator to synthesize an implementation from the module design, then the developers have to read the generated code to write the unit test cases. In this case, the readability of the generated code could become a major obstacle. Similarly, using formal models to represent system designs and module designs makes it harder for test engineers to write integration test cases and function test cases. Usually extra documentation is needed to make it work.

2.2 CBSE, MDE and Others

Component-based software engineering (CBSE) and model-driven engineering (MDE) are two distinct methodologies from a design perspective. However, in practical applications, these two approaches have a close relationship and can easily be confused. Given that this relationship is derived from the involvement of formal methods (the topic discussed in this article), we provide a brief explanation based on our view.

From a design perspective, CBSE emphasizes the relationship between the whole and its parts, with the goal of maximizing reusability. On the other hand, MDE focuses on the relationship between abstraction and concreteness (e.g., requirements and design, domain-specific and generic), with the aim of building tool-chains for analysis and verification, and reducing the complexity of software development.

In the field of model-driven development, models typically have strict formal semantics. Additionally, since abstract models often do not need to address specific details, they naturally exhibit better reusability.

Taking programming languages as an example, interfaces (in object-oriented programming languages) and function declarations (in procedure-oriented programming languages) themselves are abstractions, yet their presence allows for flexible and interchangeable implementations that serve as reusable components.

Therefore, model-driven development and component-based development are highly compatible. Conversely, once there is the involvement of strict formal semantics in the paradigm of CBSE, it naturally introduces the paradigm of model-driven development, as the relationship between components and the system itself corresponds to that of concreteness and abstraction.

Although CBSE and MDE are two different development paradigms, when formal semantics are introduced, they often converge in engineering practice. Therefore, while we discuss the challenges of component-based development, there will also be many concepts derived from the field of model-driven development in the subsequent sections.

3   Applications of Formal CBSE

The CBSE methodology has gained wide attention and application in the industry. However, the integration of formal methods and CBSE (referred as *Formal CBSE* in the following) is still limited to very specific domains. Some examples are introduced in this section.

3.1   Avionics and Railway Software

The avionics and railway software are representative sectors for safety-critical software systems. Software of this type are characterized by long development cycles, high cost of upgrades, and severe consequences in case of failures. Similar industries also include aerospace software, nuclear power control software, etc., but they will not be separately mentioned here due to their lower level of commercialization.

Avionics Software. Various tools such as SCADE (Safety Critical Application Development Environment) [7] with its model checking capabilities, Simulink and Event-B have been applied in the development of the aircraft control and display systems. For example, [18] used SCADE and its formal verification component, the Design Verifier, to assess the design correctness of a sensor voter algorithm. The algorithm is representative of embedded software used in many aircraft systems.

Railway Software. B-Method [6] and Event-B [42] have been successfully applied to modeling and code generation for subway systems. The work in [26] combines Event-B with a component-based reuse strategy realized with session types, and ensures global safety of railway interlocking systems by the local verification of entities. [32] used a subset of the SysML language for modeling of railway systems, automatically transformed the models to Event-B and finally imported the models into the RODIN platform [1] for formal verification.

3.2   Automobile Software

Vehicle software is a domain that requires a high demand for both component-based development and formal verification.

On one hand, due to the complex architecture and numerous devices (chips, peripherals, etc.) involved in the automotive industry, which have relatively stable accompanying software, it makes the field highly suitable for component-based development. The emergence of the AUTOSAR standard [21] further promotes the application of component-based development in the domain of vehicle software. The AUTOSAR standard divides vehicle software into three layers: Application Layer, Runtime Environment, and Basic Software. Each layer is further divided into several abstract components, with constraints defined in the standard. For example, the Basic Software is further divided into different layers such as Services, ECU Abstraction, Microcontroller Abstraction and Complex Drivers, and these layers are further divided into functional groups. Examples of Services include System, Memory and Communication Services. Such a layered architecture allows software and hardware suppliers to design components according to AUTOSAR specifications, ensuring that these components can be integrated together to form a complete system.

On the other hand, the automotive industry has stringent requirements for safety and reliability, creating ample opportunities for formal methods. In the functional safety standard ISO 26262, formal verification is considered an important added advantage. Currently, widely used CBSE tools that integrate formal verification capabilities include model-driven development tools like SCADE [7] and Ptolemy [10], which are based on synchronous data flow. Some modern tools targeting at automatic end-to-end compositional verification of automotive software system properties have been developed as well, such as EVA [14].

3.3   Industrial Manufacturing Software

Advanced industrial manufacturing, particularly in the modeling and analysis of control algorithms, is a typical application area for component-based development. Among this area, Matlab Simulink [19] and LabVIEW [9] are among the most general-purpose component-based development tools.

In this domain, formal methods find application mainly in the following two areas:

1. *Verification of control logics.* Large-scale industrial manufacturing processes often involve multiple concurrent control logics with complex interaction behaviors. Methods such as model checking can be employed to mitigate issues like deadlock and reduce failure rates, thereby enhancing overall stability of the manufacturing assembly line [47].
2. *Semantical analysis of control logics and algorithms.* This includes evaluating and analyzing the execution time [34] of control algorithms, stability analysis of controlled physical systems, optimization and synthesis of controllers [36], etc. By analyzing the results, algorithm optimization can be guided to improve efficiency and yield rates of the manufacturing assembly line.

It is worth noting that although Simulink has integrated basic formal verification capabilities such as Simulink Design Verifier [30] for quite some time, it is rare to see developers using them in practical applications. More often, developers choose domain-specific formal analysis tools based on the specific requirements of their scenarios, such as Coco Platform [15] and LSAT [35]. (Besides, there are much more domain-specific or proprietary formal analysis tools developed by large companies, but many of them are not publicly available.)

4 Challenges

After finishing the previous sections, attentive readers may now raise the following question:

> "The domains that have been listed above are quite familiar to the CBSE researchers. So, how to evaluate the applications of CBSE, especially equipped with formal methods, in the broader software industry?"

Objectively speaking, the CBSE methodology has already had a significant and far-reaching impact on the software industry. Derivatives such as cloud computing and micro-services architecture have emerged based on this idea [22, 28]. However, the widespread application of component-based software engineering combined with formal verification remains challenging. This section primarily focuses on the challenges in this aspect.

4.1 Hard to Keep Consistency Between Implementations and Models

Large-scale industrial software is typically developed through collaborative efforts, resulting in a fast-paced development cycle. If formal CBSE methods fail to cover the entire process ranging from design to code implementation (as mentioned in Sect. 2.1), potential inconsistencies could exist between the formal models and their corresponding implementations. For example,

- As the software evolves, new features can not be formalized using the formal model. This usually happens when the developers used a simpler formal model in order to simplify the verification.
- The benefits of formal verification have been claimed at the first time when the software is released to the customers. After that, continuous investment on formal verification does not produce comparable commercial value. Consequently, some teams choose to use formal methods in the prototype development and use traditional methods (or CBSE but without formal methods) to develop the released versions.

While the initial design models undergo comprehensive formal verification, ensuring the correctness of the software, we have observed that, as development speed accelerates, most development teams do not consistently invest manpower in model construction and verification. As a result, this eventually leads to disparities between the released software and the verified models in terms of formal semantics.

Unfortunately, this challenge is not exclusive to the application of formal verification in component-based software engineering. To the best of our knowledge, in most formal verification scenarios we are suffering from the same situation.

4.2   Lack of Life-Cycle Maintainability

When aiming to address the aforementioned issue of semantic inconsistencies, it is natural to consider the introduction of code generation tools to directly generate executable code from formally verified models. Currently, many component-based engineering tools, whether or not equipped with formal proof capabilities, support this functionality.

However, it is unfortunate that additional engineering maintainability issues are introduced by code generation. For example,

- *Poor readability of generated implementations.* Automatically generated code often lacks readability. For developers such generated codes are hard to read and comprehend. The problem becomes even worse without automated test-case generation techniques.
- *Increased integration complexity.* Industrial software projects often involve the incorporation of third-party components. It is typically difficult to invoke these third-party components from the code generated by CBSE tools. In practice, developers often need to develop compatibility layers to invoke third-party components. Conversely, the introduction of such compatibility layers and third-party libraries significantly increases the risk of inconsistencies between the models and their implementations.
- *Lengthened error-fixing cycles.* If the code is automatically generated by a tool, it implies that developers cannot manually modify the code itself (as any changes would be overwritten in the next model modification). Consequently, when clients discover bugs in the software, even minor issues require modifications starting from the models, leading to a longer overall repair process and an inability to achieve prompt customer response.

On the other hand, if any bug is found in the code generation methodology itself, it is usually hard to motivate the tool developers (community or commercial entities) to fix the bug. For developers, an unavoidable fatal error may impede the entire CBSE development process, posing significant risks to commercial software.

4.3   Fragmented Requirements from Developers

One interesting aspect of component-based development is the introduction of various development roles during the development process, each with different tool requirements. For example, in the field of control algorithms, developers of specific components (such as PID controllers, various filters, etc.) are often more concerned about their runtime efficiency and may optimize them for specific hardware mechanisms. On the other hand, the algorithm designers, who combine these components, focus more on the performance of the algorithm itself, such as convergence speed and other parameters. They also want to validate the component's parameters to prevent potential configuration errors. Developers working on higher-level business logic focus on the correctness and stability of the entire business chain, such as the presence of deadlocks or the yield rate of the production line.

Fragmented requirements often lead to two possible outcomes. Either the blind accumulation of numerous features increases the learning cost of the tool significantly, diminishing its performance and usability, or different domains start customizing the tool themselves, leading to changes in data structures due to the addition of functionalities, resulting in siloed deliveries and a fragmented tool ecosystem. Both outcomes are detrimental to the promotion of component-based development.

4.4   Extra Learning Cost for Various Tools

Introducing new tools requires extra learning cost, and due to the inherent knowledge requirements of formal methods, the learning cost can be relatively high. What's worse is that fragmented domain demands often introduce a plethora of different tools. These tools may have different input languages, output formats, and user interfaces that may vary significantly in terms of user experience. In such cases, the learning cost incurred may offset the benefits.

5   Discussion

Based on the our academic research and practical experience in component-based development in the past decades, we present some open ideas in this section and hope that they can facilitate the broader application of component-based development methods based on the formal theories in the software development domain.

5.1   LLM-Aided Explanation of Codes and Exceptions

The idea aims to tackle the challenges in Sect. 4.1 and Sect. 4.2.

To the best of our knowledge, it is difficult to find mature solutions to maintain consistency between specification models and their implementations. One typical approach in industry is to extract models from black-box implementations through model extraction, e.g. using active learning [24, 33]. However, such model extraction algorithms usually fail on complex implementations.

Program synthesis is the process of automatically generating a program or code snippet that is consistent with a given model, a formal specification or a natural language description. It involves a family of techniques such as deep reinforcement learning [44], constraint solving [16] and symbolic execution [38]. The goal of program synthesis is to minimize human intervention in the coding process, reduce program errors and complexity, and improve productivity.

Recently, the rapid evolution of Large-Language Models (LLMs [46]) brings us a new chance. Large-language models, literally, are artificial intelligence models that relies on huge parameters sets to achieve general-purpose language tasks,

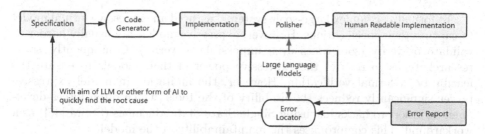

Fig. 2. Integrating LLM in CBSE Workflow

such as language understanding and generation. Modern LLMs (e.g. ChatGPT, Llama2 [40], etc.) has been fine-tuned to perform program-related tasks [40] like code generation and explanation.

On one direction, the Github Copilot [4], built by Microsoft, has already proved its ability to understand and synthesize programs. On the other direction, there are already promising research on using LLMs in reversed engineering [31]. With the help of LLMs, it would be possible to increase the readability of generated code and help the developers to quickly locate the root cause in the models when bugs are reported. A possible integration of LLM into the CBSE workflow is shown in Fig. 2.

In this example the LLM is mainly used to help with:

- *Code polishing.* Automatically generated or synthesized code are usually lack of readability. However, with the background knowledge given (domain knowledge and the algorithm of code generation, etc.), the LLM is able to understand the behavior of the code, and then perform semantically equivalent code transformation.
- *Error locating.* When errors happen in generated programs, with the background knowledge (the same as above) a LLM could be able to help locating the root cause of the error in higher level specifications. So that the developer would be able to easily fix the error by updating the specification (instead of directly updating the generated code). This is helpful to keep a short iteration cycle even in software developed by formal CBSE methodology.

5.2 Decoupling Formal Specification from Verification

This idea aims to tackle the challenges in Sect. 4.1 and Sect. 4.2. Its core motivation is to enrich the formal model so that it can cover the specifications in various domains and abstraction levels.

Formal methods are mathematically rigorous techniques for the specification, development, analysis, and verification of software and hardware systems [11] where formal specification serves as the fundamental model for the remaining parts.

To formalize complex large-scale systems, we need an expressive formal model to capture their specifications. However, expressive models are usually harder to validate or verify (sometimes even impossible to verify). Consequently, some research tends to restrict the expressive power of their models to ensure the feasibility of formal verification. However, the limitation in model expressive power significantly reduces the usability of the tools, which often forces developers to resort to hacks like using Single-State State Machines (SSSM [45]) as a workaround. This compromises the maintainability of the model.

Decoupling formal specification from formal verification means using highly expressive formal semantics in component-based development tools and integrating them throughout the entire development life-cycle. For safety-critical and reliability-critical components, we can require developers to use a verifiable subset, while allowing more flexible and complex models in other parts. For other components, we can engage other light-weighted methods such as bounded model checking, model-driven testing, and static analysis to enhance software reliability.

5.3   Layered Modeling Through Domain-Specific Languages

This idea aims to tackle the challenge in Sect. 4.3. It also provides a technical basis for the idea discussed in Sect. 5.4.

Hiding unnecessary complexity can significantly improve development efficiency and reduce potential errors. This is an undeniable truth in the field of software development. Principles such as the Dependency Inversion Principle in software architecture are aimed at hiding unnecessary complexity. Considering the example of different roles mentioned in the previous section, if we hide formal semantics, analysis, and verification features that are irrelevant to each role during component-based development, it can alleviate the issues of fragmented domain requirements and high learning costs.

To achieve this goal, introducing Domain-Specific Languages (DSLs) is undoubtedly a good approach. Compared to general-purpose programming languages, DSLs are designed specifically for certain domains and provide simple and intuitive programming languages that are only used to describe domain-specific data objects or program behavior. For example, JSON and SQL# can be considered typical DSLs for data modeling and querying [25, 29], respectively.

By using DSLs, we can layer the tool framework for component-based development, including a foundational model layer and a domain model layer. The foundational model layer consists of a more expressive formal semantic model, on which various common capabilities can be built, such as analysis, verification, and code generation. On top of the foundational model, there can be multiple modeling domains, each based on a DSL, including model transformation algorithms and a set of domain-specific functionalities. Through these algorithms, the domain models can be transformed into the foundational model, thereby reusing the capabilities of the foundational model, while also enabling the construction of domain-specific analysis capabilities. The foundational model is invisible to ordinary users. In this way, by sinking common capabilities and data structures as much as possible, this architecture can hide complexity while ensuring that fragmented domain requirements are met, as illustrated in Fig. 3.

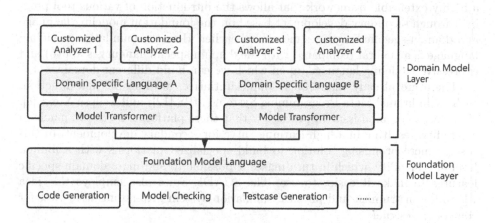

Fig. 3. Domain-specific Layered Modeling

There are quite a few previous works where domain-specific languages are applied to component-based development. For example, AADL (Architecture Analysis and Design Language) [20] supports *an annex mechanism* which can be used to extend its semantics [?]. However, such examples often focus on using DSLs to enhance the extensibility of the framework rather than hiding unnecessary complexity.

The primary purpose of using DSLs in these cases is to provide a specialized language that captures the domain-specific concepts and features required for component-based development. This allows developers to express their ideas and intentions in a more concise and natural way, improving productivity and reducing errors. DSLs can also provide higher-level abstractions that hide low-level details and technical complexities, making the development process more accessible to domain experts.

5.4   Providing Multi-Domain Integrated Development Environment

This idea aims to tackle the challenge in Sect. 4.4.

To reduce the extra learning effort required for users, a possible approach is to provide similar user experiences and logical workflows across different tools.

Taking Matlab as an example, we can use Simulink to build data flow models and Stateflow to construct state machines for controlling the data flow. This approach reduces the learning curve for users while enhancing their overall experience. SCADE takes it a step further by integrating requirements management (which can be considered as another domain, but not necessarily formal) and UI design into the same tool platform. This allows users to complete end-to-end design and implementation in certain scenarios using SCADE.

To make Formal CBSE tools applicable to a wider range of scenarios, we need a highly extensible framework that allows the introduction of various tool plug-ins through secondary development based on the foundation modeling language and domain-specific languages mentioned earlier. Here the foundation modeling language is a general (compared with the domain-specific languages above) and expressive modeling language, upon which we can build different domains.

There are already many open-source initiatives working in this direction, such as Jetbrains' Meta Programming System (MPS [12]) and Eclipse Xtext [8]. However, a common issue in practice is that these platforms focus too much on generality, resulting in a high learning curve for secondary development. In this case, a more promising solution to build specialized platforms with a limited level of scalability, which in turn makes it possible to add more domain-specific features to make it easier to use. The SCADE Suite also employs this idea where the instrument and physical simulation system give an intuitive view for controller designers.

6   Conclusion

This paper is not exactly a scientific research paper but rather resembles list of open topics to the researchers in the area of component-based software engineering, especially formal aspects of component software. We outline several typical scenarios in the industry where formal methods are used in conjunction with component-based development. It summarizes some of the challenges faced in the practical application of component-based development with formal methods and presents open viewpoints that may aid in the promotion of component-based development methods. The intention is for these challenges and perspectives to provide researchers and tool developers with insights into the current state of component-based development in industrial applications, helping them identify valuable research directions.

Acknowledgements. This research was sponsored by the National Natural Science Foundation of China under Grant No. 62172019, and CCF-Huawei Populus Grove Fund.

References

1. Abrial, J., Butler, M.J., Hallerstede, S., Hoang, T.S., Mehta, F., Voisin, L.: Rodin: an open toolset for modelling and reasoning in Event-B. Int. J. Softw. Tools Technol. Transf. **12**(6), 447–466 (2010)
2. Ahmad, E., Dong, Y., Wang, S., Zhan, N., Zou, L.: Adding formal meanings to AADL with hybrid annex. In: Lanese, I., Madelaine, E. (eds.) FACS 2014. LNCS, vol. 8997, pp. 228–247. Springer, Cham (2015). https://doi.org/10.1007/978-3-319-15317-9_15

3. Arbab, F.: Coordination for component composition. In: Liu, Z., Barbosa, L.S. (eds.) Proceedings of the International Workshop on Formal Aspects of Component Software, FACS 2005, Macao, 24–25 October 2005. Electronic Notes in Theoretical Computer Science, vol. 160, pp. 15–40. Elsevier (2005)
4. Barke, S., James, M.B., Polikarpova, N.: Grounded copilot: how programmers interact with code-generating models. Proc. ACM Program. Lang. **7**(OOPSLA1), 85–111 (2023)
5. Basu, A., et al.: Rigorous component-based system design using the BIP framework. IEEE Softw. **28**(3), 41–48 (2011)
6. Behm, P., Benoit, P., Faivre, A., Meynadier, J.-M.: Météor: a successful application of B in a large project. In: Wing, J.M., Woodcock, J., Davies, J. (eds.) FM 1999. LNCS, vol. 1708, pp. 369–387. Springer, Heidelberg (1999). https://doi.org/10.1007/3-540-48119-2_22
7. Berry, G.: Synchronous design and verification of critical embedded systems using SCADE and Esterel. In: Leue, S., Merino, P. (eds.) FMICS 2007. LNCS, vol. 4916, p. 2. Springer, Heidelberg (2008). https://doi.org/10.1007/978-3-540-79707-4_2
8. Bettini, L.: Implementing Domain-Specific Languages with Xtext and Xtend. Packt Publishing Ltd., Birmingham (2016)
9. Bitter, R., Mohiuddin, T., Nawrocki, M.: LabVIEW: Advanced Programming Techniques. CRC Press, Boca Raton (2006)
10. Buck, J.T., Ha, S., Lee, E.A., Messerschmitt, D.G.: Ptolemy: a framework for simulating and prototyping heterogenous systems. Int. J. Comput. Simul. **4**(2) (1994)
11. Butler, R.W.: What is formal methods? NASA LaRC Formal Methods Program (2001)
12. Campagne, F.: The MPS Language Workbench: Volume I, vol. 1. Fabien Campagne (2014)
13. Chouali, S., Boukerche, A., Mostefaoui, A., Merzoug, M.A.: Ensuring the compatibility of autonomous electric vehicles components through a formal approach based on interaction protocols. IEEE Trans. Veh. Technol. **72**(2), 1530–1544 (2023)
14. Cimatti, A., et al.: EVA: a tool for the compositional verification of AUTOSAR models. In: Sankaranarayanan, S., Sharygina, N. (eds.) TACAS 2023. LNCS, vol. 13994, pp. 3–10. Springer, Cham (2023). https://doi.org/10.1007/978-3-031-30820-8_1
15. Cocotec.io: Cocotec: All systems go. https://cocotec.io/
16. Colón, M.A.: Schema-guided synthesis of imperative programs by constraint solving. In: Etalle, S. (ed.) LOPSTR 2004. LNCS, vol. 3573, pp. 166–181. Springer, Heidelberg (2005). https://doi.org/10.1007/11506676_11
17. Criado, J., Asensio, J.A., Padilla, N., Iribarne, L.: Integrating cyber-physical systems in a component-based approach for smart homes. Sensors **18**(7), 2156 (2018)
18. Dajani-Brown, S., Cofer, D., Bouali, A.: Formal verification of an avionics sensor voter using SCADE. In: Lakhnech, Y., Yovine, S. (eds.) FORMATS/FTRTFT - 2004. LNCS, vol. 3253, pp. 5–20. Springer, Heidelberg (2004). https://doi.org/10.1007/978-3-540-30206-3_3
19. Simulation and model-based design (2020). https://www.mathworks.com/products/simulink.html
20. Feiler, P.H., Gluch, D.P.: Model-Based Engineering with AADL - An Introduction to the SAE Architecture Analysis and Design Language. SEI Scrics in Software Engineering. Addison-Wesley (2012)
21. Fürst, S., Bechter, M.: Autosar for connected and autonomous vehicles: the autosar adaptive platform. In: Proceedings of DSN-w 2016, pp. 215–217. IEEE (2016)

22. De Giacomo, G., Lenzerini, M., Leotta, F., Mecella, M.: From component-based architectures to microservices: a 25-years-long journey in designing and realizing service-based systems. In: Aiello, M., Bouguettaya, A., Tamburri, D.A., van den Heuvel, W.-J. (eds.) Next-Gen Digital Services. A Retrospective and Roadmap for Service Computing of the Future. LNCS, vol. 12521, pp. 3–15. Springer, Cham (2021). https://doi.org/10.1007/978-3-030-73203-5_1

23. Jifeng, H., Li, X., Liu, Z.: Component-based software engineering. In: Van Hung, D., Wirsing, M. (eds.) ICTAC 2005. LNCS, vol. 3722, pp. 70–95. Springer, Heidelberg (2005). https://doi.org/10.1007/11560647_5

24. Hendriks, D., Aslam, K.: A systematic approach for interfacing component-based software with an active automata learning tool. In: Margaria, T., Steffen, B. (eds.) ISoLA 2022. LNCS, vol. 13702, pp. 216–236. Springer, Cham (2022). https://doi.org/10.1007/978-3-031-19756-7_13

25. Hu, Y., Jiang, H., Tang, H., Lin, X., Hu, Z.: SQL#: a language for maintainable and debuggable database queries. Int. J. Softw. Eng. Knowl. Eng. **33**(5), 619–649 (2023)

26. Kiss, T., Janosi-Rancz, K.T.: Developing railway interlocking systems with session types and Event-B. In: 11th IEEE International Symposium on Applied Computational Intelligence and Informatics, SACI 2016, Timisoara, Romania, 12–14 May 2016, pp. 93–98. IEEE (2016)

27. Li, Y., Sun, M.: Component-based modeling in mediator. In: Proença, J., Lumpe, M. (eds.) FACS 2017. LNCS, vol. 10487, pp. 1–19. Springer, Cham (2017). https://doi.org/10.1007/978-3-319-68034-7_1

28. Liu, C., Yu, Q., Zhang, T., Guo, Z.: Component-based cloud computing service architecture for measurement system. In: 2013 IEEE International Conference on Green Computing and Communications (GreenCom) and IEEE Internet of Things (iThings) and IEEE Cyber, Physical and Social Computing (CPSCom), Beijing, China, 20–23 August 2013, pp. 1650–1655. IEEE (2013)

29. McNutt, A.M.: No grammar to rule them all: a survey of JSON-style DSLs for visualization. IEEE Trans. Vis. Comput. Graph. **29**(1), 160–170 (2023)

30. Miranda, B., Masini, H., Reis, R.: Using simulink design verifier for automatic generation of requirements-based tests. In: Bjørner, N., de Boer, F. (eds.) FM 2015. LNCS, vol. 9109, pp. 601–604. Springer, Cham (2015). https://doi.org/10.1007/978-3-319-19249-9_42

31. Pearce, H., Tan, B., Krishnamurthy, P., Khorrami, F., Karri, R., Dolan-Gavitt, B.: Pop quiz! can a large language model help with reverse engineering? CoRR abs/2202.01142 (2022). https://arxiv.org/abs/2202.01142

32. Salunkhe, S., Berglehner, R., Rasheeq, A.: Automatic transformation of SysML model to event-B model for railway CCS application. In: Raschke, A., Méry, D. (eds.) ABZ 2021. LNCS, vol. 12709, pp. 143–149. Springer, Cham (2021). https://doi.org/10.1007/978-3-030-77543-8_14

33. Sanchez, L., Groote, J.F., Schiffelers, R.R.H.: Active learning of industrial software with data. In: Hojjat, H., Massink, M. (eds.) FSEN 2019. LNCS, vol. 11761, pp. 95–110. Springer, Cham (2019). https://doi.org/10.1007/978-3-030-31517-7_7

34. van der Sanden, B., et al.: Compositional specification of functionality and timing of manufacturing systems. In: Drechsler, R., Wille, R. (eds.) Proceedings of FDL 2016, pp. 1–8. IEEE (2016)

35. van der Sanden, B., Blankenstein, Y., Schiffelers, R.R.H., Voeten, J.: LSAT: specification and analysis of product logistics in flexible manufacturing systems. In: Proceedings of CASE 2021, pp. 1–8. IEEE (2021)

36. van der Sanden, B., Geilen, M., Reniers, M.A., Basten, T.: Partial-order reduction for supervisory controller synthesis. IEEE Trans. Autom. Control **67**(2), 870–885 (2022)

37. Sifakis, J.: Component-based construction of real-time systems in BIP. In: Bouajjani, A., Maler, O. (eds.) CAV 2009. LNCS, vol. 5643, pp. 33–34. Springer, Heidelberg (2009). https://doi.org/10.1007/978-3-642-02658-4_4

38. Ströder, T.: Symbolic execution and program synthesis: a general methodology for software verification. Ph.D. thesis, RWTH Aachen University, Germany (2019)

39. Szyperski, C., Gruntz, D., Murer, S.: Component Software – Beyond Object-Oriented Programming, 2nd edn. Publishing House of Electronics Industry (2003)

40. Touvron, H., et al.: Llama 2: open foundation and fine-tuned chat models. CoRR abs/2307.09288 (2023). https://doi.org/10.48550/arXiv.2307.09288

41. Trivedi, P., Zulkernine, F.H.: Componentry analysis of intelligent transportation systems in smart cities towards a connected future. In: 22nd IEEE International Conference on High Performance Computing and Communications; 18th IEEE International Conference on Smart City; 6th IEEE International Conference on Data Science and Systems, HPCC/SmartCity/DSS 2020, Yanuca Island, Cuvu, Fiji, 14–16 December 2020, pp. 1073–1079. IEEE (2020)

42. Ait Wakrime, A., Ben Ayed, R., Collart-Dutilleul, S., Ledru, Y., Idani, A.: Formalizing railway signaling system ERTMS/ETCS using UML/Event-B. In: Abdelwahed, E.H., Bellatreche, L., Golfarelli, M., Méry, D., Ordonez, C. (eds.) MEDI 2018. LNCS, vol. 11163, pp. 321 330. Springer, Cham (2018). https://doi.org/10.1007/978-3-030-00856-7_21

43. Whitten, J.L., Bentley, L.D., Ho, T.I.: Systems Analysis & Design Methods. Times Mirror/Mosby College Publishing (1986)

44. Yang, M., Zhang, D.: Deep reinforcement learning guided decision tree learning for program synthesis. In: Zhang, T., Xia, X., Novielli, N. (eds.) IEEE International Conference on Software Analysis, Evolution and Reengineering, SANER 2023, Taipa, Macao, 21–24 March 2023, pp. 925–932. IEEE (2023)

45. Yang, N., Cuijpers, P.J.L., Schiffelers, R.R.H., Lukkien, J., Serebrenik, A.: Single-state state machines in model-driven software engineering: an exploratory study. Empir. Softw. Eng. **26**(6), 124 (2021)

46. Zhao, W.X., et al.: A survey of large language models. CoRR abs/2303.18223 (2023)

47. Zheng, Z., Tian, J., Zhao, T.: Refining operation guidelines with model-checking-aided FRAM to improve manufacturing processes: a case study for aeroengine blade forging. Cogn. Technol. Work **18**(4), 777–791 (2016)

Formal Aspects of Component Software

An Overview on Concepts and Relations of Different Theories

Zhiming Liu, Jiadong Teng, and Bo Liu

School of Computer and Information Science, Southwest University, Chongqing, China
liubocq@swu.edu.cn

Abstract. The *International Symposium on Formal Aspects of Component Software* (FACS) was inaugurated two decades ago in response to the major software development paradigm shift from *structured development* and *object-oriented development* to *component-based software development* (CBSD) and *service-oriented architecture* (SOA). FACS is dedicated to fostering a deeper understanding of the distinctive aspects, promoting research, education, technological advancement, and the practical application of CBSD technology. On the 20th anniversary of FACS, it is appropriate to briefly recall its background and history, thereby highlighting its contributions to the community. Taking this opportunity, we focus on the discussion to elucidate the important aspects of component software that require to be and have been considered in formal theories. Leveraging the refinement of component and object-oriented systems (rCOS) as a framework, we provide an overview of these formal theories and discuss their relationships. We intend to express a vision that different theories and methods are required for different aspects in a CBSD process, and also different formal theories are required even for a particular aspect. However, ensuring their consistent application remains a major challenge and this is a main barrier to the effective industry adoption of CBSD. Furthermore, we delineate emerging challenges and prospects associated with integrating formal methods for modelling and design human-cyber-physical systems (HCPS) - hybrid integration encompassing cyber systems, physical systems, and the mixed human and machine intelligence.

Keywords: Component-Based Software · Formal Aspects · Human-Cyber-Physical Systems · Linking Formal Methods · rCOS

1 Introduction

In general, we understand *software development* or *software engineering* as being about transforming processes carrying out by instances of domain concepts into their programming models through activities of software *requirements analysis*, *design*, *verification*, *implementation*, *deployment* and *maintenance* (now possibly better to say *evolution*). It was at the turn of the last century and this century when a major software development paradigm shift from *structured development* and *object-oriented development* to *component-based development* (CBD) and *service-oriented architecture* (SOA)

Supported by the Chinese National NSF grant (No. 62032019) and the Southwest University Research Development grant (No. SWU116007).

was happening. By then, the academic community had gained quite a mature understanding and plenty practice of *structured software development* and *object-oriented software development*. It was in that background that *International Symposium on Formal Aspects of Component Software* (FACS) started first as a workshop. In this section, we introduce the key ideas, concepts and aspects of component-based software systems and their development. As a paper for FACS anniversary track, however, we first give a brief summary of the 20 years of the workshops and conferences of FACS[1].

Table 1. 20 years of the FACS workshops and conferences

FACS Year [References]	Venue	Editors of Proceedings	Invited speakers	No. accepted papers
FACS 2022 [68]	Virtual Event	S.L.T. Tarifa and J. Proença	C. Baier, R. Neves and V. Stolz	13
FACS 2021 [65]	Virtual Event	G. Salaün and A. Wijs	R. Calinescu and C. Pasareanu	9
FACS 2020	Cancelled due to Covid Pandemic			
FACS 2019 [4]	Amsterdam, The Netherlands	F. Arbab and S-S. Jongmans	C. Ghezzi, K.G. Larsen, and W. Fokkink	14
FACS 2018 [6]	Pohang, South Korea	K. Bae and P.C. Ölveczky	E.A. Lee and G. Rosu	15
FACS 2017 [63]	Braga, Portugal	J. Proença and M. Lumpe	D. Costa and C. Palamidessi	14
FACS 2016 [30]	Besançon, France	O. Kouchnarenko, R. Khosravi	H. Giese and K-K. Lau	16
FACS 2015 [9]	Niterói, Brazil	C. Braga and P.C. Ölveczky	M. Wirsing, D. Déharbe, and R. Cerqueira	17
FACS 2014 [32]	Bertinoro, Italy	I. Lanese and E. Madelaine	H. Veith, R. de Nicola, and J-B. Stefani	22
FACS 2013 [19]	Nanchang, China	J.L. Fiadeiro, Z. Liu and J. Xue	C. Zhou, A. Legay and J. Misra	22
FACS 2012 [62]	Mountain View, CA, USA	C.S. Păsăreanu and G. Salaün	T. Bultan and S. Qadeer	16
FACS 2011 [5]	Oslo, Norway	F. Arbab and P.C. Ölveczky	J. Meseguer, J. Rushby, and K. Stølen	21
FACS 2010 [8]	Guimarães, Portugal	L.S. Barbosa and M. Lumpe	S. Seshia and L. Caires	20
FACS 2009 [57]	Eindhoven, the Netherlands	M. Sun, B. Schätz	G. Döhmen and J. Rutten	12
FACS 2008 [12]	Malaga, Spain	C. Canal, C.S. Pasareanu	Lack of data	13
FACS 2007 [52]	Sophia-Antipolis, France	M. Lumpe, E. Madelaine	C. Pasareanu and E. Zimeo	17
FACS 2006 [56]	Prague, Czech Republic	V. Mencl, F.S. de Boer	P. Van Roy and D. Caromel	16
FACS 2005 [38]	Macao, China	Z. Liu, L.S. Barbosa	F. Arbab, P. Ciancarini, J. He and R. Hennicker	22
FACS 2003 [44]	Pisa, Italy,	Z. Liu and He J	M. Broy and T. Maibaum	11

1.1 International Symposium on Formal Aspects of Component Software

The first FACS workshop was an one-day event associated with the *12th International Symposium of Formal Methods Europe* (FME), now called *International Symposium of Formal Methods* (FM), which was held September 8–14, 2003 in Pisa of Italy. The purpose of the workshop was to promote understanding of the software paradigm shift and to explore how formal methods can augment understanding of component-based technology, thereby encouraging further research and educational pursuits in formal methods and tools for the component-based construction of software systems. The program consisted of two invited talks by Manfred Broy and Tom Maibaum and a few regular presentations. An edited volume "Mathematical Frameworks for Component Software" [44] was then published with 11 chapters, including the refined versions of the invited talks, presentations at the workshop, and some other papers submitted in response to the call for contributions of the volume. The papers focus on mathematical models that identify the "core" concepts as their first-class modelling elements, including *interfaces*, *contracts*, connectors, and *services*. Each chapter provides a clear definition of components, articulated through a set of key aspects, and discusses challenges

---

[1] The first author of this paper is the founder of FACS.

related to the specification and verification of these aspects. Moreover, some papers delve into issues concerning the refinement, composition, coordination, and orchestration of software components in both individual software components and broader component-based software system development.

Originally, there were no intention to establish the FACS workshop as a recurring annual event. However, the overwhelming expression of interest led to its organisation again in 2005. Consequently, FACS workshop series have been conducted annually, with proceedings being published in Springer Electronic Notes in Theoretical Computer Science (ENTCS). In 2010, the FACS workshop series evolved into an international symposium. Subsequently, the proceedings have been documented in Springer Lecture Notes in Computer Sciences, and select papers have been chosen for publication in scholarly journals, predominantly in the Science of Computer Programming (JSCP).

Over the last two decades, FACS workshops and conferences have made good contributions to advancing research, education, and application in component-based software technology. The summary presented in Table 1 alongside the papers featured in the proceedings highlight the attraction of these FACS events for a number of excellent scientists and researchers who have showcased their work. Acknowledgements are extended to colleagues who have served as members of the steering committee, event organisers, members of the program committees, the reviewers, and editors of the proceedings. The utmost appreciation is reserved for the authors for their invaluable contributions. In this paper, we endeavour to cite proceedings from FACS workshops and conferences wherever possible to demonstrate their far-reaching impact.

Throughout the past 20 years, FACS workshops and conferences FACS have made their contributions to promote the research, education, and application in component-based software technology. The summary in Table 1 and the papers in the proceedings show that the FACS events have attracted a large number of excellent scientists and researchers to present their work there. We would like to pay tributes to the colleagues who have served as members of the steering committee, organisers of the events, members of the program committees, the reviewers and editors of the proceedings, and most of all to the authors, for their contributions. In this paper, we cite proceedings of FACS workshops and conferences wherever we can with the intention to show their impact.

1.2   Component-Based Software Development

Component-based software or *component software* generally refers to software constructed from individual components. This idea has been present since the advent of assembly programming, which is perceived as the craft of assembling instructions or "components" to create programs. Wheeler's subroutines, also known as Wheeler jumping routines [69], can be considered as sizable, reusable "components" of that era[2]. Subsequent to this, more generalised abstractions of functions and procedures were implemented in high-level programming languages such as FORTRAN and PASCAL.

It is widely agreed that the idea of developing software systems by using available software components was first proposed by Douglas McIlroy in his invited talk, *mass-produced software components* [55], during the NATO conference on software

---

[2] At a time when there is lack of hardware support to remember the return address of the routine.

engineering. The conference was held in Garmisch, Germany, in 1968, to address the problem of "software crisis". There, he called for the development of a "software component industry" to produce components which can be used in different jobs of software development.

The philosophy of constructing large software systems from components is also foundational to *Structured Programming* [18]. It is the cornerstone of object-oriented programming [20,58] too. Indeed, it is reasonable to affirm that every programming language incorporates some form of abstraction mechanism that facilitates the design and composition of components.

Thus, researchers in CBSD frequently find themselves addressing the question of *whether CBSD distinctly differs from structured and object-oriented development*. In other words, answering the following questions:

1. What characteristics distinguish components in CBSD from program instructions, routines, functions, classes, objects, libraries of classes or routines, and packages?
2. What distinctions exist between the "components", "composition", and "refinement" of components in CBSD and the analogous "components", "composition", and "refinement" of programs in structured and object-oriented programming?

In this paper, our goal is to demonstrate how research on the formal aspects of component software can elucidate the aforementioned differences and relations. To accomplish this, we employ the rCOS method as the framework to identify various aspects of software components, component-based software systems, and their development, illustrating the methods to formalise, refine, decompose, and compose them. This allows for a clear comparison with the facets of structured and object-oriented software systems. Moreover, we argue for the need for various formal theories for modelling different aspects, and even different formal theories for the same aspects. We engage in a discussion on the interoperable use of these diverse formal theories. In addition, we outline research challenges and opportunities in component-based modelling and design of human-cyber-physical systems (HCPSs), which are networked systems with mixed intelligence.

1.3   Organisation

In Sect. 2, we introduce the key characteristics of software components and present the calculi of *designs* and *reactive designs* as preliminaries. These will serve as the semantic basis for our discussions in subsequent sections. Section 3 follows, defining interfaces and contracts. The main content is housed in Sect. 4, where we provide an overview of formal models of aspects and discuss their support for the separation of concerns in CBSE. In Sect. 5, we outline the processes of software component development and system development in CBSE, discussing the principal shortcomings in the current state of the art and challenges to industrial adoption. In Sect. 6, we propose a future research direction for CBSE-specifically, component-based design and evolution of human-cyber-physical systems (HCPSs)-and discuss related challenges. We draw conclusions at the end of this section.

2 Preliminaries

In this section, we discuss the key characteristics of software components and introduce the basics of *design calculus* [46], which will serve as the semantic basis for linking different formal theories and techniques in modelling and verification of various aspects of component-based software.

2.1 Characteristics of Software Components

To address the two questions raised at the end of the previous section, we consider the definition of software component given by Szyperski in his book [66]:

> A software component is a unit of composition with contractually specified interfaces and explicit context dependencies only. A software component can be deployed independently and is subject to composition by third parties.

This definition characterises the requirements for a "component" to be *reusable* in different software systems, which is also the major concern expressed in McIlroy's talk [55]. As highlighted in [28], the concept of reusability defined herein contrasts with the "reusability" achieved through generalisation in object-oriented programming. The latter necessitates the "rewriting" of classes (either generalised or specialised) being reused. The discussion in paper [28] extends further to apply the above four criteria to discern whether elements like an assembly language instruction, a library routine, a class in a library, or a package can qualify as software components. In this context, it is concluded that an assembly language instruction does not qualify as a component due to its inability to be independently deployed. Library routines and classes may be considered components, but a package, lacking an interface, does not meet the criteria for a component. However, this reasoning is somewhat unconvincing, primarily due to a lack of rigour in both syntax and semantics.

It is important to note that the four characteristics must be described in the models of different kind of requirements of components. Each kind of requirements is called an *aspect* [28]. Components of different types of software systems have different aspects - be they sequential, concurrent and distributed, real-time, or embedded systems. Questions are raised, for instance, about the aspects that need specification in the interface contract for a component within concurrent or embedded systems.

In this paper, we take the above four basic characteristics of components as the basis and describe them in a formal model of interface contracts. An interface contract of a component is defined in the design calculus of UTP [46] and it can be factorised into specifications of different aspects. We show how different aspects embraced in an interface contract of a component can be refined in different stages of the component development, from the requirements, through the design, to coding and evolution.

2.2 Designs

We use the notion of contracts to characterise the different aspects of components. There are different theories for different aspects, and a single aspect can also be modelled using various theories. For instance, when developing reactive components, Hoare logic can be applied to specify functionality, the theory of I/O automata can model reactive behaviuor, and sequence diagram notation can represent interactions with the environment. Nonetheless, Hoare logic is often used in conjunction with traces for interactions to compose reactive components. Moreover, models or specifications, whether addressing identical or differing aspects, must maintain consistent relations. Hence, there's a requirement for a model that facilitates the definition and verification of their consistency. To provide a formal definition of contracts and formalise various aspects, we first introduce the notations of *design* and *reactive designs*. These are deployed to define the meanings of sequential and reactive programs [22,23,46], respectively.

Definition 1 (Design). *Given a set of state variables X, a design D over X is first-order predicate of the form $p \vdash Q$ with free variables in $X \cup \{ok\}$ and $X' = \{x' \mid x \in X\} \cup \{ok'\}$ The semantics of the design is defined as the implication $(ok \wedge p) \rightarrow Q \wedge ok'$, where*

- *a variable $x \in X$ represents the initial value of x in the state when the program starts to execute, and the $x' \in X'$ the value of x in the state when the execution terminates;*
- *p is predicate which is called the **precondition** and it only refers to free variables in X, and Q the **postcondition** and only refers to free variablesin X and X'; and*
- *ok and ok' are Boolean variables representing the observations about if the execution terminates, and the program is in a terminated state when ok is true, and the execution terminates when ok' is true.*

The set X of variables is called the **alphabet** of the design D. A *state* of D is a mapping s which assigns each variable x in X a value an ok a Boolean value. The precondition p and postcondition Q of a design are interpreted on the set of pairs (s, s') of states. The meaning $(ok \wedge p) \rightarrow Q \wedge ok'$ of $p \vdash Q$ says that if the execution starts well in a state s (i.e. ok is *true* in s) in which the precondition p holds, the execution will terminates in a state s' (i.e. ok' is *true* in s) such that the postcondition Q holds for (s, s').

Let $X = \{x, y, z\}$ be a set of integer variables, the design

$$x > 0 \vdash (x' = (x - 1)) \wedge (y' = x) \wedge (z' = z)$$

specifies a program such that

- from any state s in which the value of x is positive, i.e. $s(x) > 0$, its execution terminates and the execution decrements the initial value x by 1, assigns y to the initial value of x, and does not change the initial value of z, i.e. the execution terminates in a state s' such that $(s'(x) = s(x) - 1) \wedge (s'(y) = s(x)) \wedge (s'(z) = s(z))$; and
- from any state s in which the value of x is not positive, i.e. $s(x) \leq 0$, any execution is possible (acceptable), including non-terminating execution.

Program **if** $(x > 0)$ **then** $(y := x; x := x - 1)$ correctly implements this specification. Notice the difference between this design from the design below

$$x > 0 \vdash x' = (x - 1) \wedge (y' = x')$$

Note that the set of designs over X form a proper subset of first-order predicates with free variables in X, X' and $\{ok, ok'\}$. We thus require that the set of designs is closed to the common structure operations of programs, such as *assignments*, *sequencing*, *choice*, and *iteration*. To this end, we define the following primitives and operations on the set $\mathcal{P}_X = \{D \mid D \text{ is a design over } X\}$

- *skip*: $\mathbf{skip} \overset{def}{=} true \vdash \wedge_{x \in X}(x' = x)$, that is, *program* $\mathbf{skip}$ *always terminates but its execution does not change any variables*;
- *chaotic program*: $\mathbf{chaos} \overset{def}{=} fale \vdash true$, that is, *program* $\mathbf{chaos}$ *behaves as chaos and may exhibit any behaviour*;
- **assignment**: $x := e \overset{def}{=} true \vdash (x' = e) \wedge_{y \in X - \{x\}} (y' = y)$, namely, *this assignment always terminates, and its execution only changes the variable x to the value of expression x obtained in the initial, keeping the other variables unchanged* (it has no side effect);
- **sequencing**: $(D_1; D_2) \overset{def}{=} \exists(X_o, b).(D_1[X_o/X', b/ok'] \wedge D_1[X_o/X, b/ok])$, that is, *this sequence statement first executes D_1 and reaches a state (X_o, b) and then it executes D_2 from this state*;
- **conditional choice**: $D_1 \lhd B \rhd D_2 \overset{def}{=} (B \wedge D_1) \vee (\neg B \wedge D_2)$, that is, *this conditional choice statement behaves like D_1 if B is evaluated to true in the initial state, it behaves like D_2, otherwise*;
- **loop**: $(B * D) \overset{def}{=} \mu\mathcal{X}.(D; \mathcal{X}) \lhd B \rhd \mathbf{skip}$, that is, *the loop statement is the least fixed point of the recursion $\mathcal{X} = (D; \mathcal{X}) \lhd B \rhd \mathbf{skip}$.*

The above formalisation is given based on the following assumptions.

1. For the assignment, the type of x and that of e are *compatible*, and the value of e in the initial state is defined. These conditions can be explicitly expressed in precondition. We omit them for simplicity.
2. Similarly, the expression B in the conditional choice and loop is also assumed to be evaluated as a Boolean value.

For the sequence statement, b is either *true* or *false*. In former case, the execution of D_1 terminates well and execution of D_2 starts well (i.e. $ok = true$), and in the latter case the execution of D_1 is chaotic, and so is the whole sequence statement.

It is important to note that the existence of (least) fixed-point for the loop is proved based on the theorem that the partial order $(\mathcal{P}, \sqsubseteq)$, forms a *complete lattice*. In the lattice, the partial order $\sqsubseteq$ of the lattice is called the *refinement relation* among designs, and it is defined as $D_1 \sqsubseteq D_2$ if $D_2 \rightarrow D_1$ is valid. The bottom element is $\mathbf{chaos}$ and the top element is the *angelic design* defined by $true \vdash false$, denoted as $\mathbf{angel}$. Thus, per Tarski's fixed-point theorem, the least fixed point of the loop indeed exists. We proceed to present the ensuing theorem about designs [46].

Theorem 1. *The set $\mathcal{P}_X$ of designs over X is closed with respect to the program structure operations. That is, given any two design D, D_1 and D_2 and any Boolean expression B such that B only mentions variables in X, the formulas defined for skip, chaos, assignment, sequencing, conditional choice and loop are also designs.*

We can add the non-deterministic choice operation $(D_1 \sqcap D_2)$, which is defined by $(D_1 \vee D_2)$, for nondeterministic programming language.

We notice that while both the left zero law $((\mathbf{chaos}; D) = \mathbf{chaos})$ and the left unit law $((\mathbf{skip}, D) = D)$ hold for designs, neither the right zero law $((D; \mathbf{chaos}) = \mathbf{chaos})$ nor the right unit law $((D; \mathbf{skip}) = D)$ generally apply. We enforce these laws as *healthiness conditions* on the set of designs [46]. Furthermore, the assumption of side effect freedom for assignments in OO programming is generally not valid. In rCOS, we have developed a calculus of designs tailored for OO programming languages [23].

The design calculus is used to define the semantics of imperative (or procedural) programming languages, which are utilised for the implementation of components within the paradigm of structured software engineering. Meanwhile, the OO extension of the calculus in rCOS [23] can be employed to define the semantics of OO programming languages for implementing components within an OO software development paradigm [17]. Nonetheless, components implemented using these languages lack abstractions for concurrency and synchronisation.

2.3 Reactive Designs

To model communication and synchronisation, we define the notion of *reactive design*. This is done by incorporating two fresh Boolean variables, *wait* and *wait'*, to represent *observables of synchronisation*. A design D on an alphabet X with observables $\{ok, ok', wait, wait'\}$ is called a reactive design if it satisfies the healthiness condition $\mathcal{W}(D) = D$, where the transformation $\mathcal{W}$ is defined as follows:

$$\mathcal{W}(D) \overset{def}{=} (\mathit{true} \vdash \mathit{wait'} \wedge ((X' = X) \wedge ok = ok'))) \triangleleft \mathit{wait} \triangleright D$$

To specify a reactive program, it's necessary to specify the synchronisation conditions of an activity with other activities, as well as the functionality of the activities. To this end, we introduce the concept of *guarded designs* represented as $g \& D$, where D is a design, and g is a Boolean expression of the given X, called the *guard* of the design. The semantics of $g \& D$ is defined as $D \triangleleft g \triangleright (\mathit{true} \vdash \mathit{wait'} \wedge (X' = X) \wedge (ok = ok'))$, where X is the alphabet of the design D.

It is straightforward to prove that the transformation $\mathcal{W}$ is monotonic (with respect to the implication), thereby establishing that the set of reactive designs constitutes a complete lattice. Moreover, the subsequent properties are valid for reactive designs.

Theorem 2. *For a given set of variables X,*

(1) For any design D, $\mathcal{W}^2(D) = \mathcal{W}(D)$;
(2) $\mathcal{W}$ is (nearly) closed for sequencing: $(\mathcal{W}(D_1); \mathcal{W})(D_2) = \mathcal{W}(D_1; \mathcal{W}(D_2))$.
(3) $\mathcal{W}$ is closed for non-deterministic choice: $(\mathcal{W}(D_1) \vee \mathcal{W}(D_2)) = \mathcal{W}(D_1 \vee D_2)$;
(4) $\mathcal{W}$ is closed for conditional choice: $(\mathcal{W}(D_1) \triangleleft b \triangleright \mathcal{W}(D_2)) = \mathcal{W}(D_1 \triangleleft b \triangleright D_2)$.

Note that "$=$" stands for logical equivalence. Furthermore, the following properties hold for guarded designs.

Theorem 3. *For a given set of variables X and if D, D_1 and D_2 are reactive designs,*

1. *$g\&D$ is a reactive design;*
2. *$(g\&D_1 \vee g\&D_2) = g\&(D_1 \vee D_2)$;*
3. *$(g_1\&D_1 \lhd B \rhd g_2\&D_2) = (g_1 \lhd b \rhd g_2)\&(D_1 \lhd B \rhd D_2)$;*
4. *$(g_1\&D_1; g_2\&D_2) = g_1\&(D_1; g_2\&D_2)$.*

These properties, together with the subsequent transformations for programming commands, enable us to define a reactive program c as reactive designs of the form $g\&\mathcal{W}(c)$, where c is defined as follows:

$$c ::= \mathbf{skip} \mid \mathbf{stop} \mid \mathbf{chaos} \mid x := e \mid c \lhd B \rhd c \mid B * c$$

And

$$\mathcal{W}(\mathbf{skip}) \stackrel{def}{=} \mathcal{W}(true \vdash \neg wait' \wedge (X' = X))$$

$$\mathcal{W}(\mathbf{stop}) \stackrel{def}{=} \mathcal{W}(true \vdash wait' \wedge (X' = X))$$

$$\mathcal{W}(\mathbf{chaos}) \stackrel{def}{=} \mathcal{W}(fale \vdash true)$$

$$\mathcal{W}(x := e) \stackrel{def}{=} \mathcal{W}(true \vdash \neg wait' \wedge (x' = e) \wedge \bigwedge_{y \in X - \{x\}} (y' = y))$$

$$\mathcal{W}(c_1 \lhd B \rhd c_2) \stackrel{def}{=} \mathcal{W}(\mathcal{W}(c_1) \lhd B \rhd \mathcal{W}(c_2))$$

$$\mathcal{W}(B * c) = \mu.\mathcal{X}.(\mathcal{W}(c); \mathcal{X}) \lhd B \rhd \mathcal{W}(\mathbf{skip})$$

3   Interfaces and Contracts

We hold the view that there should be few restrictions on what qualifies as a component. A component can be, for instance, a few lines of code for a function, procedures, a service provider, a substantial component for internet search or managing databases, or a reactive system. Regardless, they should be deployable for execution and ought to have specified interfaces, which serve as the sole interaction points with the environment.

Interfaces are most important for components and their composition. Thus, we define them as the first-class model elements in our rCOS modelling theory [22,41].

Definition 2 (Interfaces). *An **interface** is a triple $\mathcal{I} = (T, X, O)$, where T is a set of type definitions, X a set of state variables with their types defined in T, and O is a set of operation signatures of the form $m(T_1\ x; T_2\ y)$ with an input parameter of type T_1 and an output parameter of type T_2.*

We permit T to encompass a set of class definitions in an OO programming language, facilitating the object-oriented implementation of components. Either the input parameter x or the output parameter y, can be vectors, possibly empty. The set X of variables denotes the state encapsulated in the component and can also be empty. A component is termed stateless when X is empty and stateful otherwise. The names in an interface are designated as the *syntactic aspects* of types, states, and functionality. These names must be derived from predefined formal languages. For instance, elements of T are names of defined types within a type theory; elements in X are well-defined identifiers in a specified programming language, and elements in O are signatures from an Interface Definition Language (IDL). These stipulations ensure syntactic consistency and well-formedness checking.

It is important to note that a component can have multiple interfaces. In rCOS, we handle this requirement at the syntactic level by defining "merging" or "union" of interfaces [22]. We now give the definition of interface contracts.

Definition 3 (Contracts). *A* **contract** *is a tuple* $C = (\mathcal{I}, \theta, \Phi)$, *where*

- $\mathcal{I} = (T, X, O)$ *is an interface,*
- θ *is a first order predicate on X which is the set of state variables of interface $\mathcal{I}$ specifying the allowable initial states of the program, called the* **initial condition,**
- Φ *is a mapping which assigns each interface operation* $m(T_1\ x; T_2\ :\ y) \in O$ *a guarded deign with input alphabet $X \cup x$ and output alphabet $X' \cup y'$.*

We observe that the contract of an interface generally defines a model for concurrent (or reactive) programs. Such models can be specified in a well-established formal theory, for instance, in the *temporal logic of actions* (TLA) [31] as a *normal form* $\theta \wedge \Box[\bigvee_{m() \in O} \Phi(m())]_X$. A well-established formal theory for concurrent programs, like TLA, facilitates the development of concurrent systems from scratch. Other renowned theories in this domain include UNITY [13], Event-B [1], and the stream calculus [10]. Such as theory can be used for the development of concurrent systems from scratch by going through a whole process of specification, decomposition, composition and verification. However, these theories often do not provide explicit support for the separation of concerns or for *black-box* integration (whether composition or assembly). Moreover, event-based theories, such as those based on *input/output automata* [54], CSP [27,64], CCS [59], and other process algebras [7], predominantly emphasize interaction and concurrency aspects.

We believe that these limitations play a significant role in the industry's lukewarm adoption of component-based development technologies. More effort should be dedicated to creating integrated development environments (IDEs) that support consistent use of various theories, techniques, and tools for diverse aspects. In the next section, we identify these aspects and identify the formal theories for their specification, decomposition, verification, and refinement.

4   Models of Aspects of Contracts

To address the aforementioned limitations, component-based development should ideally support the composition of components based on interface specifications in a black-box manner. Furthermore, it should enable an interface model to be divided into models addressing different aspects, thereby facilitating the separation of concerns in the development process. We will now delve into such a factorisation, drawing upon the interface model defined in the preceding section.

In general, a formal theory of each aspect is a mathematical logic system or a universal many-sorted algebra, which consists of a *formal language*, its *semantics* (interpretations), and a proof system defined by *axioms* and *inference rules*. We now discuss the aspects interface contracts and their formal theories.

4.1 Type Systems

Definitions 2&3 show that an interface contract is based on a *type system*. This represents a formal theory essential for addressing the *data aspect*, ensuring consistency checks for data representations of states in X, signatures of operations, and the expressions in guards and designs. The type system for an interface contract can be distinct from the type systems in various programming languages utilised for the component implementations. Nonetheless, transformations must be defined to bridge the abstract type system at the interface level with those in the programming languages. Established theories of *abstract data types* (ADT) and *algebraic specification*, such as [11], serve this purpose effectively.

Here we highlight the rCOS method [23,29,43,72], which provides a semantic theory for OO programs, encompassing an object-oriented type system. This theory extends the design calculus, offering the subsequent features:

- Variables are categorised as *public*, *protected*, or *private*. The types of variable values can either be *primitive types* or *classes*.
- An object is recursively defined as a graph structure. Nodes represent objects, while directed edges, labelled by the attribute names in the source node, denote references from one node to another (akin to a UML object diagram). Such a graph contains a unique root node, symbolising the current object.
- The system supports type casting through dynamic type binding of method invocations and type safety analysis.
- At any given execution moment, the program's state is an object graph termed the *state graph*. This represents the main program's object, implying the root is the main class's object. The object nodes within this state graph encapsulate the dynamic object types.
- Program command execution transitions from one state to another by either adding a new object to the graph (e.g., opening a new account in either a small or large bank system), altering attribute values of certain graph objects (like the *transfer()* action in a banking system), or modifying graph edges (such as enabling customer access to a shared account). Thus, the command's semantics (including method invocations) is expressed as a relation between states in UTP design form.
- To bolster incremental program development, a class declaration is also specified as a design, capturing changes in the program's static class structure. This can be seen as a textual formalisation of a UML class diagram. The class declaration is an action undertaken prior to program compilation.

Drawing from this semantic theory, *OO refinement* is articulated across three dimensions: *refinement of commands*, which encompasses method invocations, *refinement of classes* (also dubbed *OO structure refinement*), and *refinement of programs*. Class refinement also signifies *sub-typing*. Program refinement includes both the extension and modification of the program's class declarations, coupled with the refinement of the main method and other class methods. This research is elaborated in the paper [23]. The rCOS theory of OO semantics is also applicable to components within an OO programming paradigm. The soundness and (relative) completeness of the OO refinement are proven in [72].

4.2 Functionality and Synchronisation Behaviour

This model of interface contracts defines the requirements of a component, including both functional requirements and synchronisations conditions of each individual interface operation. From an operational perspective, we can divide this model into three parts:

- $\mathcal{R}$: Represents the *functional aspects* of the contract. It is a mapping that assigns a design (not a reactive design) to each operation $m()$.
- $\mathcal{S}$: Denotes a labelled state transition system. Using symbolic states, it characterises the reactive behaviour.
- $\mathcal{A}$: Captures the *data state aspects* of the contract. It is a mapping that associates each state s of the transition statement with a (or a set of) predicate formula $\mathcal{A}(s)$.

$\mathcal{R}$, $\mathcal{A}$ and the state transition system are related such that for any state transition $s \xrightarrow{m()} s'$ from s to s' by an operation $m()$ of the interface $\mathcal{I}$, it is required that the guard g holds in s and that design $\Phi(m())$ holds for (s, s'). Here, we say that g holds in s if $\mathcal{A}(s) \rightarrow g$, and (s, s') holds for $\Phi(m())$, if $p \rightarrow \mathcal{A}(s)$ and $\mathcal{A}(s') \rightarrow Q$, where p and Q are the precondition and postcondition of $\Phi(m())$, respectively.

Therefore, the functional aspect is specified, analysed, refined, and verified within its own theory, be it Hoare Logic [26], the design calculus in the Unifying Theory of Programming (UTP) [46], or the rCOS OO design calculus in [23]. On the other hand, the synchronisation aspect is modelled, designed, refined and verified using its dedicated theory, such as labelled state transition systems, interface automata [2], or UML state diagrams. It is important to note that there is not an automatic method (neither an algorithm nor syntactic rules) to factor a given contract. Typically, we construct the models for these aspects using a use-case driven approach, then merge them into a contract using the syntactic rules presented in [14], followed by a consistency check (refer to Sect. 5).

4.3 Interaction Protocols

When assembling components, checking interaction compatibility between them often benefits from a declarative protocol specification based on a set of allowable traces rather than an operational model. Consider a buffer with a capacity of one: it only permits interaction traces that alternate between *put*() and *get*(), beginning with *put*(). More generally, a buffer with capacity n would accept traces that start with *put*(). In any prefix of such traces, the number of *get*() occurrences should not exceed that of *put*(), and the count of *put*() should not be greater than n times the number of *get*().

This aspect can be addressed using the theory of regular expressions, process calculi, or sequence charts.

Often, we can express the specification of an interface contract in the simplified form $(\mathcal{I}, \theta, \mathcal{R}, \mathcal{T})$, where:

- $\mathcal{I}$ represents the interface,
- θ denotes the initial condition,
- $\mathcal{R}$ assigns a design to each operation signature $m()$ for its functionality aspect, and
- $\mathcal{T}$ is a set of traces of the form $\langle ?m_1(x_1), \ldots, ?m_k(x_k) \rangle$, representing the interaction protocol between the contract and its environment. Here, $?m_i(x_{i_j})$ signifies an invocation event of service m_i with the input value x_i.

One might wonder how such a specification relates to a contract defined in Definition 3. To our knowledge, no transformation directly from such a specification to a contract has been defined. However, one can envision a design of the component by initially designing the functionality of services without synchronisation control, followed by the design of the interaction protocol for the constraints of the invocations. Subsequently, we define the semantics of the design in terms of reactive designs. Essentially, the environment should not be blocked if it invokes the provider services following any of the traces in $\mathcal{T}$, and the execution of the invocation should be correct with respect to the functionality aspect $\mathcal{R}$. Several well-established theories support the specification and design of interaction protocols, including process calculi and sequence charts.

The main theme we propose here is the separation of the functionality and synchronisation aspects of communications. This separation enables the flexible combination of a simpler theory of functional design and component development, such as the rCOS relational calculus of components [42], with theories for component coordination, e.g. [3], and component orchestration, e.g. [60], for system assembly.

4.4 Dynamic Behaviour

A component interacts with its environment through its interfaces, which serve as access points. Its behaviour can be described by potential sequences of alternating events of invocations and returns of provided operations (or services). It's often necessary to specify potential failures that can occur during execution, namely, *deadlock* and *livelock* (or *divergence*). In this context, we define the *dynamic behaviour* of a contract $\mathcal{C}$, as outlined in Definition 3, as a pair of sets $\mathcal{B}_\mathcal{C} = (\mathcal{D}_\mathcal{C}, \mathcal{F}_\mathcal{C})$ representing *divergences* and *failures*. In this section, we provide only informal definitions of $\mathcal{D}_\mathcal{C}$ and $\mathcal{F}_\mathcal{C}$, and direct the reader to the paper [22] for a more formal discussion.

Definition 4 (Divergences). *Given a contract $\mathcal{C}$, the set $\mathcal{D}_\mathcal{C}$ of **divergences** for $\mathcal{C}$ consists of sequences of invocations $\langle ?m_1(x_1)!m_1(y_1) \ldots ?m_k(x_k)!m_k(y_k) \rangle$ of provided operations $m_i()$. These sequences end in a divergent state, meaning the execution of $\theta; m_1(x_1; y_1); \ldots; m_i(x_i; y_i)$ in a prefix of the sequence from an initial state results in ok' being false. Here, x_i and y_i represent the actual input and output parameters of the invocation $m_i()$, respectively.*

Notice that each pair of events $?m_i(x_i)!m_i(y_i)$ in the sequence corresponds the execution of an method invocation $m_i(x_i; y_i)$.

Definition 5 (Failures). *Given a contract* C, *the set* $\mathcal{F}_C$ *of* **failures** *for* C *consists of pairs* (tr, M). *Here,* tr *is a finite sequence of invocations* $\langle ?m_1(x_1)!m_1(y_1)\ldots\rangle$ *of provided operations, and* M *is a set of service invocations. One of the following conditions must hold:*

(1) tr *is the empty sequence, and* M *consists of invocations* $m(v)$ *for which the guards of their designs are false in the initial condition* θ.

(2) tr *is a trace* $\langle ?m_1(x_1)!m_1(y_1)\ldots?m_k(x_k)!m_k(y_k)\rangle$, *and* M *includes the invocations* $m(v)$ *such that after executing the invocations* $m_1(x_1, y_1)\ldots m_k(x_k, y_k)$ *from an initial state, the guard of* $m(v)$ *is false, i.e.,* $\theta; m_1(x_1, y_1); \ldots; m_k(x_k, y_k)$ *implies* $\neg g'$ *for the guard* g *of the design of* $m()$.

(3) tr *is a trace* $\langle ?m_1(x_1)!m_1(y_1)\ldots?m_k(x_k)\rangle$ *where the execution of the invocation* $m_k(v)$ *has not yet delivered an output, and* M *includes all invocations.*

(4) tr *is a trace* $\langle ?m_1(x_1)!m_1(y_1)\ldots?m_k(x_k)\rangle$, *and the execution of the invocation* $m_k(v)$ *has entered a wait state, and* M *comprises all invocations.*

(5) tr *is a divergence in* $\mathcal{D}_C$, *and* M *contains all invocations (all invocations can be refused after the execution diverges).*

We now establish the relationship between a protocol T defined in the previous subsection, and the dynamic behavior. This, in turn, relates to the contract defined in Definition 3. First, we define the set of traces for a contract C based on its failures.

$$Trace(C) \stackrel{def}{=} \{tr \mid \text{there exists a } M \text{ such that } (tr, M) \in \mathcal{F}_C\}$$

Then, a protocol T is a subset of $Prot(C) \stackrel{def}{=} \{tr \downarrow_? \mid tr \in Trace(C)\}$.

Definition 6 (Consistent Protocol of Contract). *A protocol* T *is* **consistent** *with* $(\mathcal{D}_C, \mathcal{F}_C)$ *(and hence with contract* C*), if executing any prefix of any invocation sequence* sq *in* T *does not lead to a state where all invocations to the provided services are rejected. That is, for any* $sq \in T$ *and any* $(tr, M) \in \mathcal{F}_C$ *such that* $sq = tr \downarrow_?$, *we must have* $M \neq \{m(v) \mid m() \in O\}$ *if* $tr \downarrow_?$.

We have the following theorem for the consistency between protocols and contracts.

Theorem 4. *Given a contract* C *and its protocols* T_1 *and* T_2, *we have:*

(1) *If* T_1 *is consistent with* C *and* $T_2 \subseteq T_1$, T_2 *is consistent with* C.

(2) *If both* T_1 *and* T_2 *are consistent with* C, *so is* $T_1 \cup T_2$.

(3) *If* $C_1 = (I, \theta_1, \Phi_1)$ *is another contract of interface* I *and* $\theta \sqsubseteq \theta_1$ *and* $\Phi(m()) \sqsubseteq \Phi_1(m())$ *for every operation* $m()$ *of the interface, then* T_1 *is consistent with* C_1 *if it is consistent with* C.

It is important to note that a *largest protocol* (or *weakest protocol*) of a contract exists and can be derived from the contract. We refer to [22] for technical details. The importance of the notion of consistency and its properties lies in ensuring the correct use of components in various interaction environments. Furthermore, this allows for the design of functional aspects and the tailoring of communication protocols (i.e., coordination and orchestration) to be separated.

In rCOS [16, 17], we demonstrated how protocols are specified using sequence diagrams, with their semantics defined in terms of CSP [35]; dynamic behaviour is modelled using UML state diagrams, and their semantics is also defined in terms of the failure-divergence semantics of CSP [64]. As a result, consistency can be automatically checked [48]. However, with UTP as its semantic foundation, other theories for different aspects can be utilised and their integration can be defined in the model of contracts defined in Definition 3. The sustainability of components is supported by the refinement calculus of contracts.

Definition 7 (Contract refinement). *A contract C_1 is* **refined** *by a contract C_2, denoted as $C_1 \sqsubseteq C_2$, if they have same interface, and*

(1) C_2 is not more likely to diverge, i.e. $\mathcal{D}_{C_2} \subseteq \mathcal{D}_{C_1}$; and
(2) C_2 is not more likely to block the environment, i.e. $\mathcal{F}_{C_2} \subseteq \mathcal{D}_{C_1}$.

There is an effective and complete method to prove that one contract refines another using *downward simulation* and *upward simulation*. We refer to our paper [22] for technical details.

4.5 Other Aspects

There are other aspects of components, including real-time and security, and even power consumption. In theory, these can be incrementally integrated into the above framework. For example, time can be represented using the timed sequence in RT-UML, timed transition systems [24], or timers or clocks defined in TLA [45]. We also believe that component-based architecture is quite effective for imposing various security policies, e.g. [47]. However, we leave these topics out of this paper due to the page limit.

5 Component-Based Development

It is widely agreed that component-based system development comprises both the creation of individual components and the system development using these available components. Components are expected to be heterogeneous, indicating that they can be software developed within various paradigms (potentially by different teams using distinct programming languages). Moreover, components can be hierarchical, suggesting that they can be assembled from other components. We assume that a component developer can construct hierarchical components, but the system developer might not necessarily be aware of these hierarchies. Nevertheless, the development process of component-based systems does involve specifying the architecture for component composition or integration. It is worth noting that the composition language (e.g., Orc [60] or Reo [3]) might differ from the languages utilized in component development.

It is a common contention that the system development process should be bottom-up, with the system assembled from pre-implemented or even pre-deployed components. However, such a stance is somewhat idealistic and not always practical, given the ever-evolving nature of applications. Additionally, we assert that there is always a necessity for a system requirements specification when developing a new system, even if it is based on an existing one. Moreover, there's a need to design a system architecture model derived from this requirements specification. This design would facilitate the identification of existing components, their interfaces, and protocols, ideally within a component repository equipped with management tool support. A significant challenge lies in the identification process. This involves mapping naming schemes from the requirements model to the architecture model and subsequently to pre-implemented components, such as those in component repositories. Arguably, this challenge is the most significant barrier to the effective industrial adoption of component-based technology.

Nevertheless, rCOS offers a framework for OO component-based software development, emphasising use case-driven system requirements specifications. From these specifications, a model of the system architecture is derived. Each use case is articulated as an interface contract, and the relationships among use cases, represented in UML use-case diagrams, are formalised as dependencies between interfaces. This forms an initial model of the system architecture. We will now provide an overview of this development process. For a more detailed discussion, we refer readers to our papers [17,37].

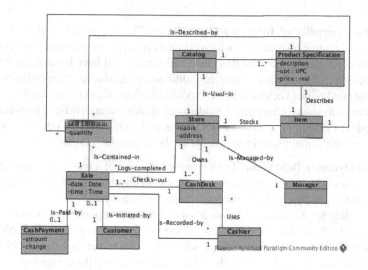

Fig. 1. An example of a conceptual class diagram

5.1 Use Case as Components

A requirements model comprises a set of interrelated use cases identified from the application domain. Each use case is modeled as an interface contract for a component. In the following discussion, we will briefly outline the steps to create a model of a use case as an interface contract. For technical details, readers are referred to rCOS-related papers, specifically the two cited here [15,17].

Identify and Represent the Interface of a Use Case. As demonstrated in the papers, an informal description of the use case is first provided using structured natural language. The operations of the provided interface for the component corresponding to a use case consist of interactions with the actors, and their symbolic names are then assigned. State variables serve as symbolic representations of the objects that must be recorded, checked, and updated, in accordance with the need to know policy [33, 36].

Identify Classes and Their Associations. The use case describe a domain process and the description involves domain concepts and objects, which the use cane needs to record, check, modify and communicates. Then an initial class diagram can then be constructed by giving symbolic names to this concepts and their relations. Figure 1 is the initial class diagram for use case "Process Sale" of the CoCoMe Example [25, 50].

Representing the Interaction Protocol. An interaction of an actor and the component is the possible events with which the actor triggers the system for the execution of an interface action. The interaction protocol between the actors and the component is represented by a sequence diagram and it is created based on the description of the use-case. Figure 2 is the initial sequence diagram use case "Process Sale" of the CoCoMe Example.

Specify Functionality of Interface Operations. The functionality of each interface operation is specified by a design, which consists of a pair of pre- and post-conditions. These conditions can initially be described informally and later formalised for consistency checking and validation. The preconditions emphasise the properties of existing objects that need to be checked, while the postconditions focus on describing what new objects are created, what changes are made to attributes, what new links between objects are formed. For examples informal descriptions of functionality of interface operations and their formalisation, we refer the reader to the papers [17, 25, 50].

Represent Dynamic Behaviour. The dynamic behavior of a use case is modeled using a state diagram. This model aids in the verification of system properties, including both safety and liveness, by employing model checking techniques.

The models for the aforementioned aspects of a component, corresponding to a use case, collectively form the interface contract for that use case. Completion and consistency checks must be conducted within the framework of the rCOS interface contracts discussed in Sect. 3. With the interface contracts of the components of the use cases, we can create the UML component diagram as the diagrammatic representation of the system architecture at the requirements modelling phase.

5.2   Component Development Process

We propose that the design process for a component begins with an OO design approach and concludes by transforming the OO design models of certain components into component-based design models. Due to space limitations, we will only outline the steps of the OO design process for a component, illustrating what models should be produced:

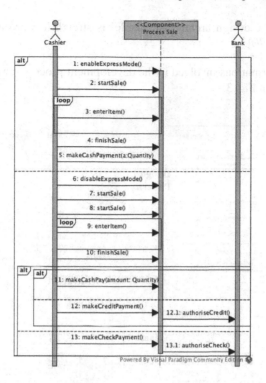

Fig. 2. An example of use-case sequence diagram

1. Begin the process by taking each use-case component and designing each of its provided operations according to its pre- and post-conditions using the OO refinement rules, especially the four patterns of GRASP [17,33].
2. Decompose the functionality of each use-case operation into internal object interaction and computation, refining the use-case sequence diagram into an object sequence diagram for the use-case [17,35].
3. During the decomposition of the functionality of use-case operations to internal object interaction and computation, refine the requirements class model into a design class model. This involves adding methods and visibilities in classes based on responsibility assignments and direction of method invocations [17].
4. Select some of the objects in an object sequence diagram as candidate component controllers. These candidates should pass an automatic check ensuring they meet six given invariant properties. Following this, transform the design sequence diagram into a component-sequence diagram [34].
5. Generate a component diagram for each use-case (this can be done automatically). This diagram should depict a decomposition of the use-case component from the requirements model into a composition of sub-components. From this, the complete component-based architecture model at the requirements level is decomposed into a component-based design architecture model [34].

6. Coding from the design architecture model is straightforward and can largely be automated [51,70].

The model transformations involved in the development process for a use-case component are depicted in Fig. 3.

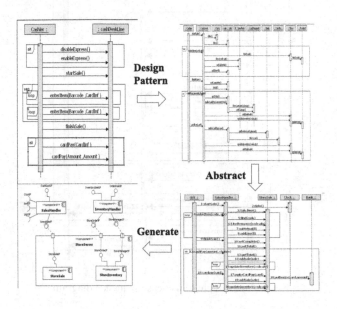

Design Pattern

Abstract

Generate

Fig. 3. Transformations from requirements to design of a component

5.3 System Development

For a given application domain, we assume a repository exists containing implemented components for a multitude of use cases, their contract specifications, information on context dependencies, and possibly their sub-components[3].

Broadly speaking, system development begins with the creation of a requirements model based on contract use cases. These use case contracts are then refined and/or decomposed into component compositions to form a system architecture model. From this point, we search the repository for candidate components that could match a component in the architecture and verify if their contracts are refinements of the component's contract within the architecture. The checks for functional requirements and synchronization requirements can be performed separately. Additionally, they can be refined individually by adding connectors and coordinators, respectively.

It is possible that for some component contracts within the architecture, there are no suitable components that can be easily adapted for implementation. In such cases, we

---

[3] However, it should be noted that we are not aware of such an existing repository.

must use the method of component development discussed in the previous subsection. The primary features of component and system development in rCOS are shown in Fig. 4.

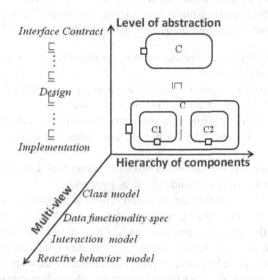

Fig. 4. Features of the rCOS modelling and development

Note that domain knowledge is crucial for providing the requirements model in terms of use cases, designing the architecture, and mapping them to components in the repository. The primary challenge in formalising the mapping and developing tool support stems from the varying naming schemes used in the requirements models, the architectural design, and the representations of the component models in the repository. In our opinion, significant effort is required in this area.

6   Future Development and Conclusion

From the discussions in the previous sections, it's evident that the modelling theories and design methods for transitional component-based software development are relatively mature. However, there are significant shortcomings in bridging the models in requirements, design, and those utilised in component repositories. These gaps are primary sources of challenges for industry adoption.

At the same time, computer systems are rapidly evolving to be more networked, hybrid, and larger in scale. Consequently, their software systems also become more intricate. Consider scenarios where components can be cyber-physical system (CPS) devices, artificial neural networks, or even humans. These systems are widely called *Human-Cyber-Physical Systems* (HCPS).

6.1 Extend rCOS for Model Human-Cyber-Physical Systems

We are currently working on a project titled *Theory of Human-Cyber-Physical Computing and Software Defined Methodology*. Our aim is to extend the rCOS component-based modelling notation to model software architectures of HCPS. We view the architecture of an HCPS as comprising *cyber systems*, *communication networks*, *physical processes*, *human processes*, and *interfaces*. Specifically:

- Physical processes can encompass various operations, such as mechanical, electrical, and chemical processes.
- Cyber systems (or information systems) are computing entities. Within this:
  - Some are dedicated to data collection and processing.
  - Others, termed controlling systems, are responsible for making control decisions based on the data provided by the aforementioned systems to manage physical processes.
- Human processes involve making control decisions based on information received from information systems to guide physical processes.
- Interfaces serve as middleware between physical systems, cyber systems, and humans. These include sensors, actuators, A/C and C/A converters, among others.
- Sensors detect the physical processes, gather data about the behavior of these processes, and relay this data to the information systems via the network.
- Control decisions made by both computers and humans are dispatched as commands through the network to the appropriate actuators, which then execute the corresponding control actions.

There is a need for system software to coordinate and orchestrate the behaviors of the component systems, as well as for scheduling the physical, network, hardware, human, and software resources. Specifically, components that can facilitate the switching of control between human and computer controllers are crucial.

Our primary challenge is to extend the rCOS contracts of interfaces to accommodate *cyber-physical interfaces* (CP-interfaces) or *hybrid interfaces*. A CP-interface incorporates field variables that include both *signals*-representing information about the states of physical processes-and program variables. It encompasses both program operations and signals for interaction with the component's environment. Additionally, a signal can be either discrete or continuous. A contract for a CP-interface comprises a provided CP interface, a required CP interface, and a specification detailing the functionality of the program operations as well as the behaviour (expressed through differential or difference equations) of the signals in the interfaces.

We suggest defining the dynamic behavior of such a contract using two hybrid input/output automata [53]. These include one for the provided CP interface and another for the required CP interface. They should be specifiable in Hybrid CSP [21] and analysable using Hybrid Hoare logic [73]. Our preliminary ideas on this extension can be found in [39,40,49], with a proof-of-concept example provided in [61].

A significant challenge in modelling HCPS is the absence of a computational model and theory for human interactions with cyber and physical systems. We propose a model of *human-cyber-physical automata* (HCPA). In this model, human behaviour is represented by a neural network, and the controller responsible for control switching between human and machine is depicted as an oracle with an associated learning model. Importantly, we are not aiming to model generalised human intelligence but rather the behaviour of a human in a specific application when executing their tasks. Researching a comprehensive theory will require addressing the complexities of integrating traditional computational models with AI models. We have given an initial definition to HCPA that includes only one human process to control a physical process in tandem with digital controllers. This model, along with a proof-of-concept case study, is presented in the extended abstract of the invited talk [71]. A full and extended version of this work is now available in the paper [67]. For further exploration of the research challenges in this project, we direct readers to the editorial paper [49] and the lecture notes available at [39]. The research will significantly encompass the controllability and composability of AI systems, their integration with traditional computational systems, and the reliability of these hybrid systems.

6.2 Conclusions

To commemorate the 20th anniversary of FACS, we aim to elucidate the meaning of "formal aspects of component software". In doing so, we provide an overview of the formal models and methods that can be used for different aspects of software components and component-based software systems. The central theme we wish to emphasise is, however, that the engineering principles of separation of concerns and divide and conquer necessitate the consistent application of various theories and methods tailored to distinct aspects. Using the rCOS framework as an example, we demonstrate these theories and methods, illustrating how and when they are applied consistently throughout development. From our discussion, it is evident that while the concepts and theories are robust, there remains a gap in the evolution of engineering techniques and tool support. Moreover, we recognise that research in CBSE must advance in tandem with developments in computer-based systems, notably in areas like human-cyber-physical systems (HCPS).

References

1. Abrial, J.R.: Modeling in Event-B: System and Software Engineering. Cambridge University Press, Cambridge (2010)
2. de Alfaro, L., Henzinger, T.A.: Interface automata. SIGSOFT Softw. Eng. Notes **26**(5), 109–120 (2001)
3. Arbab, F.: Coordinated composition of software components. In: Liu, Z., He, J. (eds.) Mathematical Frameworks for Component Software, pp. 35–68. World Scientific (2006)
4. Arbab, F., Jongmans, S.-S. (eds.): FACS 2019. LNCS, vol. 12018. Springer, Cham (2020). https://doi.org/10.1007/978-3-030-40914-2
5. Arbab, F., Ölveczky, P.C. (eds.): FACS 2011. LNCS, vol. 7253. Springer, Heidelberg (2012). https://doi.org/10.1007/978-3-642-35743-5
6. Bae, K., Ölveczky, P.C. (eds.): FACS 2018. LNCS, vol. 11222. Springer, Cham (2018). https://doi.org/10.1007/978-3-030-02146-7
7. Baeten, J.C.M., Bravetti, M.: A generic process algebra. In: Algebraic Process Calculi: The First Twenty Five Years and Beyond. BRICS Notes Series NS-05-3 (2005)

8. Barbosa, L.S., Lumpe, M. (eds.): FACS 2010. LNCS, vol. 6921. Springer, Heidelberg (2012). https://doi.org/10.1007/978-3-642-27269-1

9. Braga, C., Ölveczky, P.C. (eds.): FACS 2015. LNCS, vol. 9539. Springer, Cham (2016). https://doi.org/10.1007/978-3-319-28934-2

10. Broy, M.: A theory for requirements specifications and architecture design. In: Liu, Z., He, J. (eds.) Mathematical Frameworks for Component Software, pp. 119–154. World Scientific (2006)

11. Broy, M., Wirsing, M.: On the algebraic extensions of abstract data types. In: Díaz, J., Ramos, I. (eds.) ICFPC 1981. LNCS, vol. 107, pp. 244–251. Springer, Heidelberg (1981). https://doi.org/10.1007/3-540-10699-5_101

12. Canal, C., Pasareanu, C.S. (eds.): Proceedings of the 5th International Workshop on Formal Aspects of Component Software, FACS 2008, Malaga, Spain, 10–12 September 2008, Electronic Notes in Theoretical Computer Science, vol. 260. Elsevier (2010). https://www.sciencedirect.com/journal/electronic-notes-in-theoretical-computer-science/vol/260/suppl/C

13. Chandy, K.M., Misra, J.: Parallel Program Design: A Foundation. Addison-Wesley, Reading (1988)

14. Chen, X., Liu, Z., Mencl, V.: Separation of concerns and consistent integration in requirements modelling. In: van Leeuwen, J., Italiano, G.F., van der Hoek, W., Meinel, C., Sack, H., Plášil, F. (eds.) SOFSEM 2007. LNCS, vol. 4362, pp. 819–831. Springer, Heidelberg (2007). https://doi.org/10.1007/978-3-540-69507-3_71

15. Chen, Z., et al.: Modelling with relational calculus of object and component systems - rCOS. In: Rausch, A., Reussner, R., Mirandola, R., Plášil, F. (eds.) The Common Component Modeling Example. LNCS, vol. 5153, pp. 116–145. Springer, Heidelberg (2008). https://doi.org/10.1007/978-3-540-85289-6_6

16. Chen, Z., Li, X., Liu, Z., Stolz, V., Yang, L.: Harnessing rCOS for tool support—the CoCoME experience. In: Jones, C.B., Liu, Z., Woodcock, J. (eds.) Formal Methods and Hybrid Real-Time Systems. LNCS, vol. 4700, pp. 83–114. Springer, Heidelberg (2007). https://doi.org/10.1007/978-3-540-75221-9_5

17. Chen, Z., Liu, Z., Ravn, A.P., Stolz, V., Zhan, N.: Refinement and verification in component-based model driven design. Sci. Comput. Program. 74(4), 168–196 (2009)

18. Dahl, O., Dijkstra, E., Hoare, C.: Structured Programming. Academic Press, Cambridge (1972)

19. Fiadeiro, J.L., Liu, Z., Xue, J. (eds.): FACS 2013. LNCS, vol. 8348. Springer, Cham (2014). https://doi.org/10.1007/978-3-319-07602-7

20. Goldberg, A., Robson, D.: Smalltalk-80: The Language and Its Implementation. Addison-Wesley Longman Publishing Co. Inc., Boston (1983)

21. He, J.: From CSP to hybrid systems. In: The Proceedings of A Classical Mind: Essays in Honour of C. A. R. Hoare. Prentice-Hall (1994)

22. He, J., Li, X., Liu, Z.: A theory of reactive components. Electr. Notes Theor. Comput. Sci. 160, 173–195 (2006)

23. He, J., Liu, Z., Li, X.: rCOS: a refinement calculus of object systems. Theoret. Comput. Sci. 365(1–2), 109–142 (2006)

24. Henzinger, T.A., Manna, Z., Pnueli, A.: Temporal proof methodologies for timed transition systems. Inf. Comput. 112(2), 273–337 (1994)

25. Herold, S., et al.: CoCoME - the common component modeling example. In: Rausch, A., Reussner, R., Mirandola, R., Plášil, F. (eds.) The Common Component Modeling Example. LNCS, vol. 5153, pp. 16–53. Springer, Heidelberg (2008). https://doi.org/10.1007/978-3-540-85289-6_3

26. Hoare, C.A.R.: An axiomatic basis for computer programming. Commun. ACM 12(10), 576–580 (1969)

27. Hoare, C.A.R.: Communicating sequential processes. Commun. ACM **21**(8), 666–677 (1978)
28. Holmegaard, J.P., Knudsen, J., Makowski, P., Ravn, A.P.: Formalisization in component-based developmen. In: Liu, Z., He, J. (eds.) Mathematical Frameworks for Component Software, pp. 271–295. World Scientific (2006)
29. Ke, W., Liu, Z., Wang, S., Zhao, L.: A graph-based generic type system for object-oriented programs. Front. Comput. Sci. **7**(1), 109–134 (2013)
30. Kouchnarenko, O., Khosravi, R. (eds.): FACS 2016. LNCS, vol. 10231. Springer, Cham (2017). https://doi.org/10.1007/978-3-319-57666-4
31. Lamport, L.: The temporal logic of actions. ACM Trans. Program. Lang. Syst. **16**(3), 872–923 (1994)
32. Lanese, I., Madelaine, E. (eds.): FACS 2014. LNCS, vol. 8997. Springer, Cham (2015). https://doi.org/10.1007/978-3-319-15317-9
33. Larman, C.: Applying UML and Patterns: An Introduction to Object-Oriented Analysis and Design and the Unified Process, 2nd edn. Prentice-Hall, Upper Saddle River (2001)
34. Li, D., Li, X., Liu, Z., Stolz, V.: Interactive transformations from object-oriented models to component-based models. In: Arbab, F., Ölveczky, P.C. (eds.) FACS 2011. LNCS, vol. 7253, pp. 97–114. Springer, Heidelberg (2012). https://doi.org/10.1007/978-3-642-35743-5_7
35. Li, X., Liu, Z., He, J.: A formal semantics of UML sequence diagram. In: 15th Australian Software Engineering Conference (ASWEC 2004), 13–16 April 2004, Melbourne, Australia, pp. 168–177. IEEE Computer Society (2004)
36. Liu, Z.: Software development with UML. Technical report 259, IIST, United Nations University, P.O. Box 3058, Macao (2002)
37. Liu, Z.: Linking formal methods in software development - a reflection on the development of rCOS. In: Bowen, J., Li, Q., Xu, Q. (eds.) Theories of Programming and Formal Methods. LNCS, vol. 14080, pp. 52–84. Springer, Cham (2023). https://doi.org/10.1007/978-3-031-40436 8_3
38. Liu, Z., Barbosa, L.S. (eds.): Proceedings of the International Workshop on Formal Aspects of Component Software, FACS 2005, Macao, 24–25 October 2005. Electronic Notes in Theoretical Computer Science, vol. 160. Elsevier (2006). https://www.sciencedirect.com/journal/electronic-notes-in-theoretical-computer-science/vol/160/suppl/C
39. Liu, Z., Bowen, J.P., Liu, B., Tyszberowicz, S., Zhang, T.: Software abstractions and human-cyber-physical systems architecture modelling. In: Bowen, J.P., Liu, Z., Zhang, Z. (eds.) SETSS 2019. LNCS, vol. 12154, pp. 159–219. Springer, Cham (2020). https://doi.org/10.1007/978-3-030-55089-9_5
40. Chen, X., Liu, Z.: Towards interface-driven design of evolving component-based architectures. In: Hinchey, M.G., Bowen, J.P., Olderog, E.-R. (eds.) Provably Correct Systems. NMSSE, pp. 121–148. Springer, Cham (2017). https://doi.org/10.1007/978-3-319-48628-4_6
41. Liu, Z., Jifeng, H., Li, X.: Contract oriented development of component software. In: Levy, J.-J., Mayr, E.W., Mitchell, J.C. (eds.) TCS 2004. IIFIP, vol. 155, pp. 349–366. Springer, Boston, MA (2004). https://doi.org/10.1007/1-4020-8141-3_28
42. Liu, Z., He, J., Li, X.: rCOS: a relational calculus for components. In: Liu, Z., He, J. (eds.) Mathematical Frameworks for Component Software, pp. 207–238. World Scientific (2006)
43. Liu, Z., Jifeng, H., Li, X., Chen, Y.: A relational model for formal object-oriented requirement analysis in UML. In: Dong, J.S., Woodcock, J. (eds.) ICFEM 2003. LNCS, vol. 2885, pp. 641–664. Springer, Heidelberg (2003). https://doi.org/10.1007/978-3-540-39893-6_36
44. Liu, Z., Jifeng, H. (eds.): Mathematical Frameworks for Component Software. World Scientific (2006)
45. Liu, Z., Joseph, M.: Specification and verification of fault-tolerance, timing, and scheduling. ACM Trans. Program. Lang. Syst. **21**(1), 46–89 (1999)

46. Liu, Z., Kang, E., Zhan, N.: Composition and refinement of components. In: Butterfield, A. (ed.) Post Event Proceedings of UTP08. LNCS, vol. 5713. Springer, Berlin (2009)
47. Liu, Z., Morisset, C., Stolz, V.: A component-based access control monitor. In: Margaria, T., Steffen, B. (eds.) ISoLA 2008. CCIS, vol. 17, pp. 339–353. Springer, Heidelberg (2008). https://doi.org/10.1007/978-3-540-88479-8_24
48. Liu, Z., Stolz, V.: The rCOS method in a nutshell. In: Fitzgerald, J., Larsen, P.G., Sahara, S. (eds.) Modelling and Analysis in VDM: Proceedings of the Fourth VDM/Overture Workshop. No. CS-TR-1099 in Technical Report Series, Newcastle University (2008)
49. Liu, Z., Wang, J.: Human-cyber-physical systems: concepts, challenges, and research opportunities. Frontiers Inf. Technol. Electron. Eng. 21(11), 1535–1553 (2020)
50. Liu, Z., Zhang, Z. (eds.): SETSS 2014. LNCS, vol. 9506. Springer, Cham (2016). https://doi.org/10.1007/978-3-319-29628-9
51. Long, Q., Liu, Z., Li, X., He, J.: Consistent code generation from UML models. In: Australian Software Engineering Conference, pp. 23–30. IEEE Computer Society (2005)
52. Lumpe, M., Madelaine, E. (eds.): Proceedings of the 4th International Workshop on Formal Aspects of Component Software, FACS 2007, Sophia-Antipolis, France, 19–21 September 2007. Electronic Notes in Theoretical Computer Science, vol. 215. Elsevier (2008). https://www.sciencedirect.com/journal/electronic-notes-in-theoretical-computer-science/vol/215/suppl/C
53. Lynch, N., Segala, R., Vaandrager, F.: Hybrid I/O automata. Inf. Comput. 185, 105–157 (2003)
54. Lynch, N.A., Tuttle, M.R.: An introduction to input/output automata. CWI Q. 2(3), 219–246 (1989)
55. McIlroy, M.D.: Mass produced software components. In: Software Engineering: Report of a Conference Sponsored by the NATO Science Committee, Garmisch, Germany, 7–11 October 1968. Scientific Affairs Division, NATO (1969)
56. Mencl, V., de Boer, F.S. (eds.): Proceedings of the Third International Workshop on Formal Aspects of Component Software, FACS 2006, Prague, Czech Republic, 20–22 September 2006. Electronic Notes in Theoretical Computer Science, vol. 182. Elsevier (2007). https://www.sciencedirect.com/journal/electronic-notes-in-theoretical-computer-science/vol/182/suppl/C
57. Meng, S., Schätz, B. (eds.): Proceedings of the 6th International Workshop on Formal Aspects of Component Software, FACS@FMWeek 2009, Eindhoven, The Netherlands, 2–3 November 2009. Electronic Notes in Theoretical Computer Science, vol. 263. Elsevier (2010). https://www.sciencedirect.com/journal/electronic-notes-in-theoretical-computer-science/vol/263/suppl/C
58. Meyer, B.: Object-Oriented Software Construction, 2nd edn. Prentice Hall, Hoboken (1997)
59. Milner, R.: Communication and Concurrency. Prentice-Hall Inc., Upper Saddle River (1989)
60. Misra, J.: Orchestration. In: Fiadeiro, J.L., Liu, Z., Xue, J. (eds.) FACS 2013. LNCS, vol. 8348, pp. 5–12. Springer, Cham (2014). https://doi.org/10.1007/978-3-319-07602-7_2
61. Palomar, E., Chen, X., Liu, Z., Maharjan, S., Bowen, J.P.: Component-based modelling for scalable smart city systems interoperability: a case study on integrating energy demand response systems. Sensors 16(11), 1810 (2016)
62. Păsăreanu, C.S., Salaün, G. (eds.): FACS 2012. LNCS, vol. 7684. Springer, Heidelberg (2013). https://doi.org/10.1007/978-3-642-35861-6
63. Proença, J., Lumpe, M. (eds.): FACS 2017. LNCS, vol. 10487. Springer, Cham (2017). https://doi.org/10.1007/978-3-319-68034-7
64. Roscoe, A.W.: Theory and Practice of Concurrency. Prentice-Hall, Upper Saddle River (1997)
65. Salaün, G., Wijs, A. (eds.): FACS 2021. LNCS, vol. 13077. Springer, Cham (2021). https://doi.org/10.1007/978-3-030-90636-8

66. Szyperski, C.: Component Software: Beyond Object-Oriented Programming, 2nd edn. Addison-Wesley Longman Publishing Co. Inc., Boston (2002)
67. Tang, X., Zhang, M., Liu, W., Du, B., Liu, Z.: Towards a model of human-cyber-physical automata and a synthesis framework for control policies. J. Syst. Archit. **144**, 102989 (2023)
68. Tarifa, S.L.T., Proença, J. (eds.): FACS 2022. LNCS, vol. 13712. Springer, Cham (2022). https://doi.org/10.1007/978-3-031-20872-0
69. Wheeler, D.J.: The use of sub-routines in programmes. In: Proceedings of the 1952 ACM National Meeting, Pittsburgh, USA, p. 235. ACM (1952)
70. Yang, Y., Li, X., Ke, W., Liu, Z.: Automated prototype generation from formal requirements model. IEEE Trans. Reliab. **69**(2), 632–656 (2020)
71. Zhang, M., Liu, W., Tang, X., Du, B., Liu, Z.: Human-cyber-physical automata and their synthesis. In: Seidl, H., Liu, Z., Pasareanu, C.S. (eds.) ICTAC 2022. LNCS, vol. 13572, pp. 36–41. Springer, Cham (2022). https://doi.org/10.1007/978-3-031-17715-6_4
72. Zhao, L., Liu, X., Liu, Z., Qiu, Z.: Graph transformations for object-oriented refinement. Formal Aspects Comput. **21**(1–2), 103–131 (2009)
73. Zou, L., Zhan, N., Wang, S., Fränzle, M., Qin, S.: Verifying simulink diagrams via a hybrid hoare logic prover. In: Ernst, R., Sokolsky, O. (eds.) Proceedings of the International Conference on Embedded Software, EMSOFT 2013, Montreal, QC, Canada, 29 September–4 October 2013, pp. 9:1–9:10. IEEE (2013). https://doi.org/10.1109/EMSOFT.2013.6658587

Overview on Constrained Multiparty Synchronisation in Team Automata

José Proença(⊠) (iD)

CISTER and University of Porto, Porto, Portugal
jose.proenca@fc.up.pt

Abstract. This paper provides an overview on recent work on Team Automata, whereby a network of automata interacts by synchronising actions from multiple senders and receivers. We further revisit this notion of synchronisation in other well known concurrency models, such as Reo, BIP, Choreography Automata, and Multiparty Session Types.
We address realisability of Team Automata, i.e., how to infer a network of interacting automata from a global specification, taking into account that this realisation should satisfy exactly the same properties as the global specification. In this analysis we propose a set of interesting directions of challenges and future work in the context of Team Automata or similar concurrency models.

1 Introduction

Many different formal models for concurrent systems exist, each with its own advantages and disadvantages. This short paper provides an overview on recent work on *Team Automata* (TA), and takes a step back to relate the *constrained multiparty synchronisation* of TA with other popular models in this community studying fundamentals of component-based software.

TA were initially proposed by ter Beek et al. [9], inspired by similar approaches such as I/O automata [31] and interface automata [23], whereby a network of automata with input and output labels interacts. This interaction in TA involves *multiparty synchronisations*, whereas multiple senders and receivers can participate in a single atomic global transition. It is also *constrained* because it is parameterised by a synchronisation policy that, for each label, specifies the possible number of senders and of receivers that it can have.

Composing a network of components in a team, using constrained multiparty synchronisation, yields a new transition system whose labels are interactions consisting of (i) a message name or type, (ii) a set of components that send this message, and (iii) a set of components that receive this message. Only valid interactions are allowed, i.e., that obey the associated synchronisation policy. Many topics have been investigated in the context of TA since 2003, including security [15,17], composition and expressivity [10,13], variability [8], and compatibility of components [6,7,11,22]. Labels of early versions of TA include only message names, and were later extended to explicitly capture which components

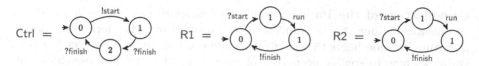

Fig. 1. Race example: a controller asks simultaneously 2 runners to start and receives a finish message once each is of them is done

participate in system transitions [11], giving rise to a concurrent semantics. This idea has been also used in Vector TA [14], where actions include active participants. In this paper we use these extended labels in TA.

We revisit several kinds of interactions in the *orchestration models* Reo [29] and BIP [20], and attempt to frame this multiparty synchronisation in the context of *choreographic models* such as choreographic automata [3] and (synchronous) multiparty session types [36]. We further address realisation of TA, i.e. how to obtain a set of interacting components from a global transition system over interactions, given the synchronisation policy. This last part is ongoing work [12], in collaboration with Rolf Hennicker and Maurice ter Beek, and include a detour on how to specify properties of interest, and how to guarantee that these are preserved when realising a global specification.

The explanations in this paper are relatively informal, driven mainly by examples. We proceed by introducing an example used throughout this paper.

Motivating Example. We use as a running example a Race system (Fig. 1), borrowed from previous work [16], consisting of 3 communicating components: a controller Ctrl and two runners R1, R2. Each is an automaton with input actions (?start and ?finish), output actions (!start and !finish), and internal actions (run). Interactions are subject to the following synchronisation policy: (i) the 3 actions start must synchronise, i.e., must be performed atomically, representing a simultaneous start of both runners; and (ii) the 3 actions finish must synchronise in pairs, i.e., each runner should atomically notify the controller that s/he finished, but the controller must receive each message one at a time. The internal action run is not involved in any interaction.

The composition of these 3 automata combined with the synchronisation restrictions of the start and finish actions yields what we call a team automaton [5, 26]. Labels of our team automaton are *interactions* involving the senders, the receivers, and the action name. E.g., "Ctrl → {R1, R2} : start" is an interaction that labels a transition in our team automaton, which follows our synchronisation policy stating that start should always have one sender and two receivers.

Many studies on TA investigate whether the components of a system will ever fail when trying to send a particular message (receptiveness), or to receive a set of possible messages (responsiveness) [6,8,10]. Another topic of interest addressed in this paper, in the context of TA or other similar systems, regards realisability or decomposition, i.e., whether these 3 components can be discovered given a description of the global behaviour capturing the same behaviour.

Organisation of the Paper. Section 2 investigates how the multiparty synchronisation of our Race example can be captured by other formalisms in the literature, considering both orchestration and choreographic languages. Section 3 addresses how to specify properties of interest for TA and proposes ideas and challenges regarding the realisation of TA. Section 4 concludes this paper.

2 Related Models

This section provides an overview of alternative approaches to model our race example with existing models that describe interaction patterns. We start with *orchestration* models, where a naive implementation of the interaction patterns leads to a centralised component controlling the interactions. We continue with *choreographic* models, which focus on how to describe the interaction *contracts* that each computational component must obey to derive an intended communication protocol, without relying on a dedicated component with the interaction logic. The precise distinction between orchestration and choreographic models is not consensual among researchers, whereas one can provide arguments why a given orchestration model can be considered to be a choreographic one and vice-versa.

Unlike the models listed below, TA identify which actions from composed components can interact by considering (1) the name of the action, (2) the direction of dataflow, and (3) the synchronisation policy (i.e., for each action, how many inputs and outputs are required or allowed). Some models below assume that all names are distinct, and that the interaction models must relate them (e.g., as an enumeration of possible sets of actions that must synchronise), sometimes leading to a more exhaustive description of the possible interactions. Other models below require interactions to be always 1-to-1 (called peer-to-peer in [9]), possibly requiring more intermediate components to capture the intended behaviour.

2.1 Orchestration

Models and languages that orchestrate components focus on how to group the interaction logic in a connector, restricting how the components can interact. This section analyses our race example in the context of the Reo [29] and the BIP [20] coordination models, since they focus on the analysis of interactions of systems with synchronous interactions that involve multiple participants. These have been compared in more detailed by Dokter et al. [24].

Reo. Reo is a graphical modelling to orchestrate components based on the synchronous composition of a set of primitive connectors [29]. Many different semantic formalisms exist to provide a semantics to Reo connectors – we focus here on constraint automata [2].

We encode our race example as a Reo connector on the left side of Fig. 2, and its semantics given by a constraint automaton on the right side. This semantics

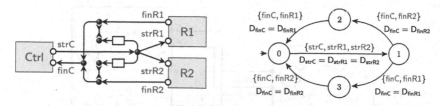

Fig. 2. Reo approach: the connector (left) built compositionally describing valid sequences of interactions from participants, and its semantics given by a constraint automata (right)

(and most Reo semantics) is stateful, i.e., the next set of possible interactions depends on the previously taken interactions. Reo is highly compositional, in the sense that the global interaction patterns result from composing simpler connectors. In our example from Fig. 2, we count 2 FIFO channels (⊡→), 2 synchronous barriers (>——<), 3 replicators (→•< after strC, finR1, and finR2), and 1 interleaving merger (>•→ before finC). Each of these has a defined semantics given by constraint automata [2], and composing these 8 automata (and hiding internal names) yields the constraint automata on the right of Fig. 2. Here each arc is labelled with a set of ports that must synchronise and a constraint over the data that must flow in each port. The direction of dataflow is not captured by constraint automata.

Reo focuses on the connector and not on the components. This means that the connector dictates which ports are active and inactive at each moment, allowing only desirable patterns of interaction. Hence, the precise behaviour of a runner, e.g., is not explicit in this model, including other internal actions that s/he may perform. When connecting a concrete runner to this connector one should verity whether it is compliant, i.e., if its possible patterns of interaction are consistent with the patterns imposed by the connector. This compliance check can be regarded as a type check against a behavioural type, as used in Session Types, described in the next section, and investigated less in the context of Reo.

BIP. BIP [20] is a language to specify the expected behaviour of components and of component architectures. A program in BIP consists of a labelled state machine for each component (**B**ehaviour), a set of possible synchronous **I**nteractions between transitions of different components, and a possible partial order among interactions (**P**riority). Semantic models, such as the algebra of connectors [20], formalise the semantics of interactions.

Our race example is modelled in BIP in Fig. 3. Similarly to TA, the components describe the core behaviour of a system, here represented as labelled transition systems (also 1-safe Petri Nets can be used in some tools). The set of valid interactions is stateless, and are depicted by connecting ports that must synchronise; the dotted connection is an alternative to the its solid counterpart, connecting R2.finish instead of R1.finish. Trigger ports, not used here, can also

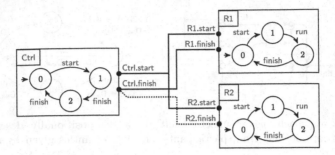

Fig. 3. BIP approach: individual components are labelled by actions, which are subject to the set of valid rendez-vous interaction constraints imposed by the middle (stateless) connector

be used instead of rendez-vous ports (●) to denote broadcasts, i.e., all siblings of a trigger in an interaction that are ready to synchronise will do so, and the trigger will always succeed.

TA share some concepts with BIP: the behaviour of the components and some synchronisation restrictions are modelled separately. Unlike TA, there is no explicit direction of dataflow. Interactions are often enriched with descriptions of how the dataflow should be updated, meaning that different connectors may treat the same ports as inputs or outputs. Unlike BIP, TA do not support any combination of synchronising ports, but only those that share the same name.

2.2   Choreographies

We consider a choreographic model to be a language or a calculus that describes the global set of valid interactions. Each interaction is typically described by (i) the name that identifies the interface, (ii) the sender, and (iii) the receiver. This richer representation of interactions, with respect to the ones used in the previous subsection, allows the behaviour of each participant to be derived from the global model, which is often the ultimate goal of these models. Most of these support only a single sender and a single receiver, and are not meant for multiparty synchronisations, although it is possible to lift many of these models to the latter case. In contrast, Reo is typically agnostic to the concrete components connected to the ports of a connector, and describes only their valid patterns of interaction, and BIP avoids specifying explicitly the global behaviour (focusing on the local behaviour).

This section provides an overview of choreographic automata [33] to illustrate an approach to reason over deterministic automata in these rich interactions, and of synchronous multiparty session types [36]. When modelling our Race example we use a variation that allows multiple senders and receivers, without providing its formal semantics.

Fig. 4. Choreographic Automaton (left), extended with multiple senders and receivers, and composed TA (right) of the automata in Fig. 1

Choreographic Automata (CA).

CA [3] are automata with labels that describe interactions, including a sender, a receiver, and a message name.

Our race example is encoded as a variation of CA on the left of Fig. 4, where we use a set of senders and receivers at each interaction to match the semantics of our composed TA. For comparison, the composed TA of our running example is depicted on the right of Fig. 4, where each state is a triple with the states of the local components, and the labels are interactions following the notation in [11]. We drop the curly brackets of singleton sets in the CA for simplicity. We also avoid representing the internal action run, although this could have been done with our extension to sets of senders and receivers, e.g., writing {R1} → {} : run to denote a send by R1 with no receivers. This would have resulted in a larger set of states capturing possible interleavings.

Multiparty Session Types (MPST).

MPST use a calculus to describe global behaviour. We can represent our race example using a variation of the global type of a synchronous MPST [28,36] as follows.

$$\lambda X \cdot \text{Ctrl} \rightarrow \{R1, R2\} : \left\{ \text{start.} \begin{pmatrix} R1 \rightarrow \text{Ctrl} : \text{finish.} \\ R2 \rightarrow \text{Ctrl} : \text{finish.} X \end{pmatrix}, \text{start.} \begin{pmatrix} R2 \rightarrow \text{Ctrl} : \text{finish.} \\ R1 \rightarrow \text{Ctrl} : \text{finish.} X \end{pmatrix} \right\}$$

This example uses a fixed point (λX) to iterate [28]. Interactions are written as {A} → {B} : {$\text{msg}_i.C_i$}$_{i \in I}$, for any sets of participants A, B and I, messages msg_i, and continuations C_i of the choreography; we omit curly brackets when there is only one element. The set of messages and continuations denotes the choices after communications.

This example does not follow the usual MPST syntax: it uses a set of receivers in the first interaction, and includes two choices of messages inside the big curly brackets that start with the same message start. Most work on MPST supports only a single sender and receiver in each interaction, and choices with syntactic restrictions, such as the need to distinguish the first message. Up to our knowledge, without these extensions and without introducing new messages the Race example cannot be faithfully modelled.

A global choreography of our reference synchronous MPST [28,36], to be valid, must obey several properties, including the need to have different messages at the start of each choice, and it must be possible to produce a projection for each participant whereas each participant does not (syntactically) distinguish choices made by other participants.

3   Behavioural Properties and Realisability

Having a TA (or another model of constrained interactions) allows us to reason over desired behavioural properties. Our properties of interest can target either a **specific** scenario, e.g., that a runner must always run before finishing, or be more **general**, e.g., no component can ever fail to send or receive messages. Hence we believe that *dynamic logic* provides the right level of abstraction to describe valid sequences of actions, which excels at capturing what actions can and cannot be performed throughout an execution. Other alternatives, such as traditional linear time logic (LTL) and computation tree logic (CTL), often focus on reachability of states and not on sequences of actions, making these less optimal for our properties of interest.

A simpler alternative to dynamic logics could be the use of regular expressions to model valid patterns of interactions. This is aligned with existing approaches such as CA that focus on the languages accepted by these automata [4] to reason over properties of such systems. The choice of using dynamic logics, regular expressions, or other model to specify properties, impacts realisability, since it is desirable for global models and their realisations to satisfy the same properties.

Many model checkers exist to verify temporal properties of concurrent systems, including mCRL2 [21] and Uppaal [18]. These support either dynamic logic (mCRL2) or CTL. Both Reo and TA have been encoded as mCRL2 processes to exploit its powerful model-checking engine [16,30,35]. The notion of synchronisation types, describing the number of possible senders and receivers by each message, can be captured using mCRL2, although it requires constructing the set of all concrete combinations of ports that can synchronise. This notion cannot directly be mapped to Uppaal, which supports only pairwise communication or a form of broadcast that differs from the one in TA (c.f. [9]).

3.1   Propositional Dynamic Logic

In our Race example it is desirable that, for any runner that starts running, it should be possible to finish her/his run. This can be expressed by the dynamic logic formula below, using regular expressions over interactions as actions, and writing *some* to denote the non-deterministic choice over any interaction:

$$\left[some^*; \mathsf{Ctrl} \to \{\mathsf{R1}, \mathsf{R2}\} : \mathsf{start} \right] \left(\begin{array}{l} \langle some^*; \mathsf{R1} \to \mathsf{Ctrl} : \mathsf{finish} \rangle \; true \; \wedge \\ \langle some^*; \mathsf{R2} \to \mathsf{Ctrl} : \mathsf{finish} \rangle \; true \end{array} \right)$$

Informally, $[\alpha]\psi$ holds if, after any of the sequence of interactions covered by the regular expression α, ψ holds. Similarly, $\langle \alpha \rangle \psi$ holds if it is possible to perform any of the sequence of interactions covered by the regular expression α, and end up in a state where ψ holds. The formula above means that, after any sequence of interactions that ends in $\mathsf{Ctrl} \to \{\mathsf{R1}, \mathsf{R2}\} : \mathsf{start}$, there must exist either a sequence of interactions ending in $\mathsf{R1} \to \mathsf{Ctrl} : \mathsf{finish}$ or in $\mathsf{R2} \to \mathsf{Ctrl} : \mathsf{finish}$.

These formulas can be expressed by the modal (μ-calculus) logic used by mCRL2. Furthermore, we exploited in recent work [16] how to generate, given a set of components of a team automaton and its synchronisation policies, both a mCRL2 model and a set of formulas that can guarantee progress. I.e., that any component who want to send a message can do so, and any component that is ready to receive messages can receive at least one of them. These two concepts are known in the literature as receptiveness and responsiveness [6], respectively.

3.2 Realisability Challenges

Realising a global specification means inferring the local behaviour of each of the involved participants, under some assumptions regarding their communication channels. We claim that realisations should satisfy the same properties of their original global models. Hence we should rely on bisimulation to compare their behaviour when using dynamic logic, since satisfaction of (propositional) dynamic logic formulas is invariant under bisimulation [19]. The choice of a different logic would lead to different equivalence notions.

Realisability has been extensively studied in the context of MPST, often guided by strict syntactic restrictions over the global protocol. These restrictions facilitate the process of building the local behaviour of each participant, by projecting the interactions to each of these participants, and provide computationally simpler mechanisms to guarantee correctness of the realisation.

Realisability has also been studied for CA [4] and for TA [13]; in both cases addressing language equivalence rather than bisimulation equivalence when comparing behaviours. In CA [4] building a local behaviour for a given participant P means producing an automaton that uses only interactions in which P is involved, such that the language is the same as the language of the global model restricted to these interactions. For instance, it is enough to hide interactions in which P does not appear (replacing them by τ), and minimise the automata collapsing these transitions. In TA [13] the language of a TA is characterised by the so-called synchronised shuffling of the behaviour of its components. Doing so, however, does not guarantee in neither cases that the realisation will satisfy the same (dynamic) logical formulas as the global protocol. As a matter of fact, some formulas in dynamic logic may satisfy either the global protocol or the composed system (exclusively) if these are language equivalent but not bisimilar.

This leads us to our ongoing effort to calculate a realisation from a global behaviour [12], i.e., from the semantics of TA with interactions as labels. Unlike the work on MPST, we try to avoid imposing syntactic restrictions on our global model, allowing our starting point to be any transition system labelled by interactions with multiple participants, instead of targeting a more practical subclass of global models. And unlike the work on CA, we try to guarantee that the realisations are bisimilar to the original model.

We identify a set of challenges when reasoning over realisability of TA.

- **How rich are the local labels?** Our initial motivating example in Fig. 1 uses labels consisting of a message name and a direction (input or output). In constraint automata (Fig. 2) there is no direction (less information), and in both CA and MPST the local participants use in each label the name, the direction, and the other participants involved (more information). Hence there will be global specifications that can be realised when producing participants with richer labels, but that cannot be realised when some information is lost (such as all participants involved). Conversely, having less information in the labels enables more compact local participants, e.g., our controller does not need to distinguish the finish from any of the two runners, while the controller derived from MPST needs to consider any interleaving of these two.

– **How to specify global specifications?** Building an automaton labelled by interactions can easily become too large due to all the combinations of concurrent actions. Hence a more compact model, such as a calculus for choreographies (with multiple senders and receivers) with a parallel operator, or other useful operators, seems to be preferable. Alternatively models based on event structures [1] or Petri Nets could also provide a compact representation of concurrent actions. Regarding the latter, the *Vector Team Automata* variation of TA [14], where vectors of local labels restrict which components participate in each global label, has a composed semantics given by a form of labelled Petri Net called *Individual Token Net Controllers*.

– **Active learning?** Active learning approaches [25] try to infer the behaviour of a system by observing the actions of an input-deterministic black-box, assuming some mechanism to discover that the inferred model is good enough. Hence, a technique to infer the behaviour of local agents from traversing a (potentially large) global state could also be adapted to infer the behaviour of a set of agents from observing and reacting to ongoing interactions. This would be an alternative to produce the global state from a given model.

– **Other communication channels?** TA focus on synchronous interactions. Relaxing this to asynchronous interactions with many participants would largely increase the complexity of the realisability analysis. Furthermore, many variations are considered in the literature, usually fixed upfront. E.g., assuming a single sorted queue between each pair of participants (blocking if the first message cannot be processed), assuming there is no order on messages, assuming there is an order that gives priority to earlier messages (but allows skipping messages), and so on. Better understanding the impact of these, or even supporting the combination of these channel mechanisms, could improve the scope of applicability of existing tools and analysis.

– **How to model families of global specifications?** Variability has been studied in TA [8] and in other models such as BIP [27], Reo [34], and Petri Nets [32]. Variability in TA, BIP, and Petri Nets meant annotating transitions with conditions over a set of features that describe whether they should be included in a given configuration. However, these are not meant to describe, e.g., a family of systems with a number n of runners, for any $n > 0$. This was attempted with Reo [34], but using a complex calculus and not targetting automatic analysis. Hence finding a good model to describe variability on the number of participants, and exploit it in the analysis of TA or providing tool support, could be a good fit for TA, which already describes desired numbers of participants involved in each channel.

4 Conclusions

This paper revisits the constrained multiparty synchronisation present in Team Automata, relating it to other popular concurrency models, guided by a race example. It further addresses verification of TA via dynamic logics, and provides a direction and challenges on how to realise team automata from global specifications. By avoiding technical details and following an example-first approach, we expect this paper to be a nice introduction to concurrency models that can synchronise multiple participants, and to provide inspiration on topics and directions that we find relevant in this area.

Acknowledgments. This work was supported by the CISTER Research Unit (UIDP/UIDB/-04234/-2020), financed by National Funds through FCT/MCTES (Portuguese Foundation for Science and Technology) and by project IBEX (PTDC/CCI-COM/-4280/-2021) financed by national funds through FCT. It is also a result of the work developed under the project Route 25 (ref. TRB/2022/00061 – C645463824-00000063) funded by NextGenerationEU, within the Recovery and Resilience Plan (RRP).

References

1. Arbach, Y., Karcher, D.S., Peters, K., Nestmann, U.: Dynamic causality in event structures. Log. Methods Comput. Sci. **14**(1) (2018). https://doi.org/10.23638/LMCS-14(1:17)2018
2. Baier, C., Sirjani, M., Arbab, F., Rutten, J.J.M.M.: Modeling component connectors in Reo by constraint automata. Sci. Comput. Program. **61**(2), 75–113 (2006). https://doi.org/10.1016/j.scico.2005.10.008
3. Barbanera, F., Lanese, I., Tuosto, E.: Choreography automata. In: Bliudze, S., Bocchi, L. (eds.) COORDINATION 2020. LNCS, vol. 12134, pp. 86–106. Springer, Cham (2020). https://doi.org/10.1007/978-3-030-50029-0_6
4. Barbanera, F., Lanese, I., Tuosto, E.: Formal choreographic languages. In: ter Beek, M.H., Sirjani, M. (eds.) COORDINATION. LNCS, vol. 13271, pp. 121–139. Springer, Cham (2022). https://doi.org/10.1007/978-3-031-08143-9_8
5. ter Beek, M.H.: Team automata: a formal approach to the modeling of collaboration between system components. Ph.D. thesis, Leiden University (2003)
6. ter Beek, M.H., Carmona, J., Hennicker, R., Kleijn, J.: Communication requirements for team automata. In: Jacquet, J.-M., Massink, M. (eds.) COORDINATION 2017. LNCS, vol. 10319, pp. 256–277. Springer, Cham (2017). https://doi.org/10.1007/978-3-319-59746-1_14
7. ter Beek, M.H., Carmona, J., Kleijn, J.: Conditions for compatibility of components. In: Margaria, T., Steffen, B. (eds.) ISoLA 2016. LNCS, vol. 9952, pp. 784–805. Springer, Cham (2016). https://doi.org/10.1007/978-3-319-47166-2_55
8. ter Beek, M.H., Cledou, G., Hennicker, R., Proença, J.: Featured team automata. In: Huisman, M., Păsăreanu, C., Zhan, N. (eds.) FM 2021. LNCS, vol. 13047, pp. 483–502. Springer, Cham (2021). https://doi.org/10.1007/978-3-030-90870-6_26

9. ter Beek, M.H., Ellis, C.A., Kleijn, J., Rozenberg, G.: Synchronizations in team automata for groupware systems. Comput. Sup. Coop. Work **12**(1), 21–69 (2003). https://doi.org/10.1023/A:1022407907596

10. ter Beek, M.H., Gadducci, F., Janssens, D.: A calculus for team automata. ENTCS **195**, 41–55 (2008). https://doi.org/10.1016/j.entcs.2007.08.022

11. ter Beek, M.H., Hennicker, R., Kleijn, J.: Compositionality of safe communication in systems of team automata. In: Pun, V.K.I., Stolz, V., Simao, A. (eds.) ICTAC 2020. LNCS, vol. 12545, pp. 200–220. Springer, Cham (2020). https://doi.org/10. 1007/978-3-030-64276-1_11

12. ter Beek, M.H., Hennicker, R., Proença, J.: Realisability of global models of inter- action. In: Ábrahám, E., Dubslaff, C., Tarifa, S.L.T. (eds.) Theoretical Aspects of Computing – ICTAC 2023. LNCS, vol. 14446, pp. 236–255. Springer, Cham (2023). https://doi.org/10.1007/978-3-031-47963-2_15

13. ter Beek, M.H., Kleijn, J.: Team automata satisfying compositionality. In: Araki, K., Gnesi, S., Mandrioli, D. (eds.) FME 2003. LNCS, vol. 2805, pp. 381–400. Springer, Heidelberg (2003). https://doi.org/10.1007/978-3-540-45236-2_22

14. ter Beek, M.H., Kleijn, J.: Vector team automata. Theor. Comput. Sci. **429**, 21–29 (2012). https://doi.org/10.1016/j.tcs.2011.12.020

15. ter Beek, M.H., Lenzini, G., Petrocchi, M.: Team automata for security: a survey. Electron. Notes Theor. Comput. Sci. **128**(5), 105–119 (2005). https://doi.org/10. 1016/j.entcs.2004.11.044

16. ter Beek, M.H., Cledou, G., Hennicker, R., Proença, J.: Can we communicate? Using dynamic logic to verify team automata. In: Chechik, M., Katoen, J.P., Leucker, M. (eds.) Proceedings of the 25th International Symposium on Formal Methods (FM 2023). LNCS, vol. 14000. Springer, Cham (2023). https://doi.org/ 10.1007/978-3-031-27481-7_9

17. ter Beek, M.H., Lenzini, G., Petrocchi, M.: A team automaton scenario for the analysis of security properties of communication protocols. J. Autom. Lang. Comb. **11**(4), 345–374 (2006). https://doi.org/10.25596/jalc-2006-345

18. Behrmann, G., David, A., Larsen, K.G.: A tutorial on Uppaal. In: Bernardo, M., Corradini, F. (eds.) Formal Methods for the Design of Real-Time Systems, Interna- tional School on Formal Methods for the Design of Computer, Communication and Software Systems, SFM-RT 2004, Bertinoro, Italy, 13–18 September 2004, Revised Lectures. LNCS, vol. 3185, pp. 200–236. Springer, Cham (2004). https://doi.org/ 10.1007/978-3-540-30080-9_7

19. van Benthem, J., van Eijck, J., Stebletsova, V.: Modal logic, transition systems and processes. J. Log. Comput. **4**(5), 811–855 (1994). https://doi.org/10.1093/logcom/ 4.5.811

20. Bliudze, S., Sifakis, J.: The algebra of connectors - structuring interaction in BIP. IEEE Trans. Comput. **57**(10), 1315–1330 (2008). https://doi.org/10.1109/ TC.2008.26

21. Bunte, O., et al.: The mCRL2 toolset for analysing concurrent systems. In: TACAS. LNCS, vol. 11428, pp. 21–39. Springer, Cham (2019). https://doi.org/10.1007/978- 3-030-17465-1_2

22. Carmona, J., Kleijn, J.: Compatibility in a multi-component environment. Theor. Comput. Sci. **484**, 1–15 (2013). https://doi.org/10.1016/j.tcs.2013.03.006

23. de Alfaro, L., Henzinger, T.A.: Interface automata. In: ESEC/FSE, pp. 109–120. ACM (2001). https://doi.org/10.1145/503209.503226

24. Dokter, K., Jongmans, S., Arbab, F., Bliudze, S.: Combine and conquer: relating BIP and Reo. J. Log. Algebraic Methods Program. **86**(1), 134–156 (2017). https:// doi.org/10.1016/j.jlamp.2016.09.008

25. al Duhaiby, O., Groote, J.F.: Active learning of decomposable systems. In: Bae, K., Bianculli, D., Gnesi, S., Plat, N. (eds.) FormaliSE@ICSE 2020: 8th International Conference on Formal Methods in Software Engineering, Seoul, Republic of Korea, 13 July 2020, pp. 1–10. ACM (2020). https://doi.org/10.1145/3372020.3391560

26. Ellis, C.A.: Team automata for groupware systems. In: Proceedings of the 1st International ACM SIGGROUP Conference on Supporting Group Work (GROUP), pp. 415–424. ACM (1997). https://doi.org/10.1145/266838.267363

27. Farhat, S., Bliudze, S., Duchien, L., Kouchnarenko, O.: Toward run-time coordination of reconfiguration requests in cloud computing systems. In: Jongmans, S., Lopes, A. (eds.) Coordination Models and Languages - 25th IFIP WG 6.1 International Conference, COORDINATION 2023, Held as Part of the 18th International Federated Conference on Distributed Computing Techniques, DisCoTec 2023, Lisbon, Portugal, 19–23 June 2023, Proceedings. LNCS, vol. 13908, pp. 271–291. Springer, Cham (2023). https://doi.org/10.1007/978-3-031-35361-1_15

28. Ghilezan, S., Jaksic, S., Pantovic, J., Scalas, A., Yoshida, N.: Precise subtyping for synchronous multiparty sessions. J. Log. Algebraic Methods Program. **104**, 127–173 (2019). https://doi.org/10.1016/j.jlamp.2018.12.002

29. Jongmans, S.T.Q., Arbab, F.: Overview of thirty semantic formalisms for Reo. Sci. Ann. Comput. Sci. **22**(1), 201–251 (2012). https://doi.org/10.7561/SACS.2012.1.201

30. Kokash, N., Krause, C., de Vink, E.P.: Reo + mCRL2: a framework for model-checking dataflow in service compositions. Formal Aspects Comput. **24**(2), 187–216 (2012). https://doi.org/10.1007/s00165-011-0191-6

31. Lynch, N.A., Tuttle, M.R.: An introduction to Input/Output automata. CWI Q. **2**(3), 219–246 (1989). https://ir.cwi.nl/pub/18164

32. Muschevici, R., Proença, J., Clarke, D.: Feature Nets: behavioural modelling of software product lines. Softw. Sys. Model. **15**(4), 1181–1206 (2016). https://doi.org/10.1007/s10270-015-0475-z

33. Orlando, S., Pasquale, V.D., Barbanera, F., Lanese, I., Tuosto, E.: Corinne, a tool for choreography automata. In: Salaün, G., Wijs, A. (eds.) Formal Aspects of Component Software - 17th International Conference, FACS 2021, Virtual Event, 28–29 October 2021, Proceedings. LNCS, vol. 13077, pp. 82–92. Springer, Cham (2021). https://doi.org/10.1007/978-3-030-90636-8_5

34. Proença, J., Clarke, D.: Typed connector families. In: Braga, C., Ölveczky, P.C. (eds.) Formal Aspects of Component Software - 12th International Conference, FACS 2015, Niterói, Brazil, 14–16 October 2015, Revised Selected Papers. LNCS, vol. 9539, pp. 294–311. Springer, Cham (2015). https://doi.org/10.1007/978-3-319-28934-2_16

35. Proença, J., Madeira, A.: Taming hierarchical connectors. In: Hojjat, H., Massink, M. (eds.) Fundamentals of Software Engineering - 8th International Conference, FSEN 2019, Tehran, Iran, 1–3 May 2019, Revised Selected Papers. LNCS, vol. 11761, pp. 186–193. Springer, Cham (2019). https://doi.org/10.1007/978-3-030-31517-7_13

36. Severi, P., Dezani-Ciancaglini, M.: Observational equivalence for multiparty sessions. Fundam. Informaticae **170**(1–3), 267–305 (2019). https://doi.org/10.3233/FI-2019-1863

Embedding Formal Verification in Model-Driven Software Engineering with SLCO: An Overview

Anton Wijs[(✉)] [iD]

Eindhoven University of Technology, Eindhoven, The Netherlands
a.j.wijs@tue.nl

Abstract. In 2009, the Simple Language of Communicating Objects (SLCO) Domain-Specific Language was designed. Since then, a range of tools have been developed around this language to conduct research on a wide range of topics, all related to the construction of complex, component-based software, with formal verification being applied in every development step. In this paper, we present this range, and draw connections between the various, at first glance disparate, research results. We discuss the current status of the SLCO framework, i.e., the language in combination with the tools, and plans for future work.

Keywords: Domain-Specific Language · Model-Driven Software Engineering · formal verification · parallel software · component-based software

1 Introduction

The development of complex software, such as component-based software, is time-consuming and error-prone. One methodology aimed at making software development more transparent and efficient is *Model-Driven Software Engineering* (MDSE) [38]. In a typical MDSE workflow, software is (mostly automatically) constructed by first creating a high-level description of the system under development, by means of a *model*. Such a model is often expressed in a *Domain-Specific Language* (DSL). This initial model is subsequently gradually refined via *model transformations*, to add information to the model in a structured way, and finally, once the model is detailed enough, derive source code that implements the low-level description of the final model (see Fig. 1). Such a workflow is also used in some *low-code application development* platforms [22].

Model transformations can be viewed as artefacts that accept a model as input, and either produce a new model (model-to-model) or code (model-to-code) as output.[1] Once defined, they can be applied automatically on models. Ideally, once the initial model has been created, and the necessary model transformations identified or designed, the MDSE procedure is fully automatic, resulting in source code that exactly implements the intended functionality, or at least requires only minor manual alteration.

---

[1] Model-to-code transformations are also known as code generators.

J. Cámara and S.-S. Jongmans (Eds.): FACS 2023, LNCS 14485, pp. 206–227, 2024.
https://doi.org/10.1007/978-3-031-52183-6_11

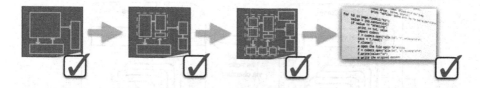

Fig. 1. Verification in a Model-Driven Software Engineering workflow.

Automatically developing software via MDSE goes a long way in reducing the introduction of errors and making software development more efficient. However, functional correctness of the developed software is not guaranteed. About 15 years ago, researchers in the Software Engineering & Technology group at the Eindhoven University of Technology, started to investigate in which ways software verification techniques could be embedded in MDSE in a seamless way [7,25]. After a collaboration with industry, it soon became clear that in order to structurally perform this research, the starting point needed to be a DSL that is relatively simple, yet expressive enough to model the basic functionality of software consisting of multiple interacting components. This lead to the creation of the *Simple Language of Communicating Objects* (SLCO) [5].

The development of SLCO and its model transformations was originally motivated by research questions addressing the internal and external quality of model transformations. The internal quality refers to the definition of a model transformation, while the external quality considers the process of applying a transformation on a model [4]. By analysing the impact of a model transformation on a given SLCO model, or the potential impact of a transformation on an arbitrary SLCO model, the external quality is assessed. Soon, however, SLCO was used for research on embedding formal verification techniques throughout the entire MDSE workflow, so that not only the initial model, but all produced artefacts can be formally verified (see the green ticks in Fig. 1).

In this paper, we present an overview of the research conducted in the last 15 years with SLCO on formal verification techniques to verify the various MDSE artefacts. While the individual results have already been published, such an overview allows viewing the bigger picture, and the directions in which the research as a whole is going in the future.

2   A DSL for Component-Based Software

In 2009, SLCO version 1.0 was developed to address a particular case study, namely the generation of Not Quite C (NQC) code for a controller of a conveyor belt built in Lego Mindstorms [6,7]. The key part of this platform is a programmable controller called RCX. It has an infrared port for communication and is connected by wires to sensors and motors for environment interaction. This imposes particular restrictions, such as the fact that communication is asynchronous and unreliable, i.e., messages may get lost. These restrictions

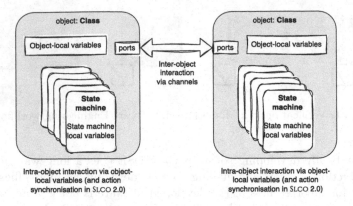

Fig. 2. The general structure of an SLCO model.

influenced the design of SLCO 1.0. For instance, besides reliable synchronous and asynchronous channels, SLCO also has an unreliable asynchronous channel as a building block. In addition, as NQC does not support arrays, these were also not added to SLCO 1.0. Finally, as NQC allows working with timers, SLCO was equipped with a delay statement, to allow expressing that a component should wait a specified number of milliseconds.

The general structure of an SLCO model is presented in Fig. 2. In a model, one or more classes are defined, which can be instantiated as objects. In a class, local variables are defined, which can be of one of the primitive types: Boolean, Integer or String. In addition, a class has a finite number of state machines. Each state machine contains a finite number of states, one of which being the *initial* state, and transitions, and optionally local variables of the primitive types. Finally, a class possibly has one or more *ports*, to which channels can be connected.

Within an object, i.e., a class instance, the state machines can interact via the variables defined at the object level. State machines in different objects can communicate via channels. In this way, SLCO can be used both for the specification of shared-memory, parallel systems and distributed systems.

Channels are connected between the ports of their respective objects. Messages sent over these channels contain a *signal*, i.e., a header, and a fixed number of values, each of a primitive type. When a channel instance is created, the type of the messages for this channel is defined, and it is specified whether the channel is synchronous or asynchronous, and in the latter case, what the size of its FIFO buffer is, and whether it is lossless or lossy.

State machines can exhibit behaviour. At any time, one of the states of a state machine is its *current* state, which initially is the initial state. If an outgoing transition of this current state is *enabled*, the state machine can fire the transition and move to the target state of that transition. With each transition, a list of zero or more *statements* are associated. If a transition has no statements, it is always enabled. If it has at least one statement, it is enabled iff its first statement is enabled. Firing a transition means considering each associated statement, in the order defined by the list, for execution. In SLCO 1.0, the following statement types are available:

- *Assignment* (x := E): assign to variable x the value defined by the expression E. Variable x can be any variable in the scope of the state machine, i.e., it can be either state machine local or local to the object containing the state machine. In expression E, references to variables in the scope of the state machine and constants can be combined with the usual operators. The expression must evaluate to a value of the same type as x. Assignments are always enabled.
- *(Boolean) Expression*: If an expression E evaluates to a Boolean value, it can be used as a stand-alone statement, and act as a guard. Such a statement is enabled iff it evaluates to **true**.
- *Send* (send <message> to <port>): send the given message via the channel connected to <port>. In case the channel is synchronous, this statement is enabled iff at least one other state machine can receive the message via the other port of the channel. If the channel is asynchronous, the statement is enabled iff the buffer of the channel is not yet full. When fired, the message is either sent to a receiving state machine (synchronous) or added to the FIFO buffer of the associated channel (asynchronous).
- *Receive* (receive <message>|<guard> from <port>): if there is a message available to be received, its signal matches the one specified, and the (Boolean expression) **guard**, which may refer to the message to be received via the variable(s) in which the message value(s) is/are to be stored, evaluates to **true**, then the receive statement is enabled, and when executed, results in the message of the sender being received (synchronous) or the first message in the FIFO buffer of the channel being received and removed from the buffer (asynchronous).
- *Delay* (after <time> ms): wait for **time** milliseconds. This statement is always enabled.

Each statement is *atomic*, i.e., its execution cannot be interrupted. Regarding concurrency, SLCO has an interleaving semantics.

The fact that a transition can have a list with more than one statement may lead to situations in which a transition is fired, but its execution cannot terminate, due to the execution reaching a statement that is not enabled. For instance, the sequence x := 0; y := 1; x = y cannot terminate, unless another state machine interferes with x and y to make the expression x = y evaluate to **true**. If the execution of a transition cannot terminate, the state machine is stuck in an intermediate state, in-between the source and target states of the fired transition. To make it simpler to reason about this, a fragment of SLCO is referred to as 'simple SLCO', which only differs from SLCO in the fact that at most one statement can be associated with each transition. A model-to-model transformation has been defined, that can transform SLCO models to semantically equivalent simple SLCO models, by introducing additional states and transitions where needed.

```
 1  model ex_chan {
 2    classes
 3      P {
 4        variables
 5          Integer x
 6        ports Out
 7        state machines
 8          SM1 {
 9            initial R0 states R1
10            transitions
11              R0 -> R1 { x = 0; x := 1 }
12              R1 -> R0 { send M(x) to Out }
13          }
14      }
15
16      Q {
17        variables
18          Integer result
19        ports In
20        state machines
21          SM2 {
22            initial S0
23            transitions
24              S0 -> S0 { receive M(result | result % 2 = 1) from In }
25          }
26      }
27    objects p: P(), q: Q()
28    channels c (Integer) sync between p.Out and q.In
29  }
```

Fig. 3. An SLCO model of a distributed system.

Finally, before we discuss the research conducted with SLCO, we present an example SLCO model in Fig. 3. The name of the model is defined at line 1. Furthermore, classes P and Q are defined at lines 3–26, and instantiated to objects p and q at line 27. In each class, variables and ports are defined (lines 4–5 and 17–18, and lines 6 and 19, respectively). Each class contains one state machine. The states of these state machines are defined at lines 9 and 22, and their transitions are listed at lines 10–12 and 23–24. Finally, a synchronous channel between the ports of objects p and q is defined at line 28. Note that q can successfully receive one message from p, as the sent message has a matching signal M and contains the value 1, which meets the requirement of q that the value must be odd. Also, only a single message can be sent, since after sending, state machine SM1 returns to state R0, at which point execution is permanently blocked, since the expression x = 0 evaluates to **false**, and once set to 1, x is never set to 0 again.

3 Verifying Model-to-Model Transformations

3.1 Reverification of Models

To investigate the ability to reason about the external quality of model-to-model transformations, a number of model-to-model transformations were developed for SLCO, using the XTEND and ATL model transformation languages and the XPAND tool [10,35], to be applied in the Lego Mindstorms conveyor belt case study. Some of these addressed refactoring aspects, such as the automatic removal of unused variables, channels and classes, changing object-local variables to state machine local ones in case they are only accessed by a single state machine, and merging objects together into a single object. The latter also affects interaction between state machines: as state machines from different objects are moved to the same object, any interaction via channels between them is transformed to interaction via shared variables. For Lego Mindstorms, this model transformation allowed to meet the criterion that the number of objects in the final model has to match the number of RCXs in the system setup.

In addition, model-to-model transformations affecting the semantics were defined, such as a transformation that introduces delays in a given set of transitions, a transformation that achieves synchronous communication with asynchronous channels, a transformation that achieves broadcasting messages between more than two state machines via a number of channels, a transformation that introduces the Alternating Bit Protocol (ABP) [26] to deal with lossy channels, and a transformation that makes the sender of a message explicit in each message.

Finally, a number of model transformations were defined to transform SLCO models to artefacts written in other languages, such as a transformation to PROMELA to allow the model checking of SLCO models with the SPIN model checker [32], a transformation to dot to allow the visualisation of state machines, and, of course, a code generator for NQC to produce source code.

The model-to-model transformation from SLCO to PROMELA was used in a first attempt to verify model-to-model transformations [6]. Since verifying the model-to-model transformations themselves would require new verification techniques, the approach was to verify a given SLCO model, and reverify it each time a model-to-model transformation that produced a refined SLCO model had been applied to it. This approach does not verify that a given model-to-model transformation is guaranteed to produce correct models in general, but at least it allows to verify that it works correctly on a case-by-case basis. After every transformation application, the resulting SLCO model was transformed to a PROMELA model, to be verified with SPIN.

However, a major drawback of this method is its limited scalability. Table 1 shows the impact of model-to-model transformations on the size of the state space of a simple SLCO model with a producer and a consumer object, i.e., a model very similar to the one of Fig. 3, in which one state machine sends messages and another one receives those messages [6,25]. Changing the initially synchronous channels to asynchronous ones doubles the size of the state space, but making this channel lossy, and introducing the ABP protocol has a significant impact on the state space size. Finally, adding delays to the transitions further increases the state space by a factor 10. Considering that this is only a model with a single channel, one can imagine the impact on models with many more channels. Because of this state space explosion, and the fact that model-to-model transformations by themselves are actually relatively small and typically only impact a particular part of a model, the ambition was soon formulated to conduct

Table 1. State space sizes of models specifying a producer and a consumer.

Model	# States	# Transitions
Synchronous	4	6
Asynchronous	8	11
Lossy + ABP	114,388	596,367
Delays	1,009,856	5,902,673

research on verifying model-to-model transformation definitions themselves, and reason about their impact on models in general.

3.2   Direct Verification of Model-to-Model Transformations

In 2011, research was started on directly verifying the impact of model-to-model transformations on models in general [60,61,64]. Contrary to the majority of the work on model transformation verification at that moment [3,55], which focussed on wellformedness of transformations, i.e., that model transformations produce syntactically correct output, and questions such as whether a model transformation is terminating and/or confluent, we decided to focus on the semantical guarantees that model-to-model transformations can provide.[2] In particular, since in model checking, models are checked w.r.t. given functional properties formalised in temporal logic, we were interested in verifying whether model-to-model transformations preserve those properties. Being able to conclude this would mean that reverification of models would no longer be needed.

Inspired by action-based model checking, we decided to reason about the semantics of SLCO models by means of *Labelled Transition Systems* (LTSs), as defined in Definition 1.

Definition 1 (Labelled Transition System). *A Labelled Transition System L is a tuple* $\langle S, A, T, \hat{s} \rangle$, *with*

- *S a finite set of states;*
- *A a set of actions;*
- $T \subseteq S \times A \times S$ *a transition relation;*
- $\hat{s} \in S$ *the initial state.*

SLCO has a formal semantics, and it was straightforward to map that semantics to LTSs. Actually, since model-to-model transformations for component-based systems tend to transform individual components, we reasoned about the semantics of component-based systems by means of *LTS networks* [41]. In such a network, the potential behaviour of each individual component is represented by an LTS, and the potential interaction between these components is defined

---

[2] Other works addressing the semantical impact of transformations include [28,33,46].

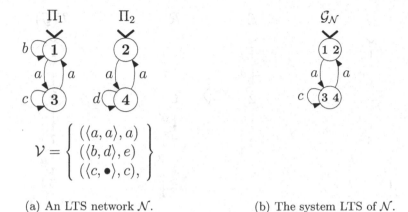

(a) An LTS network $\mathcal{N}$.   (b) The system LTS of $\mathcal{N}$.

Fig. 4. An example LTS network and its corresponding system LTS.

by a set of *synchronisation laws* $\mathcal{V}$ that expresses which actions of the individual
LTSs need to synchronise with each other, and which do not. An LTS network
semantically corresponds with an individual *system LTS*, in which the potential
behaviour of the components is interleaved, and synchronisations are applied
where needed. Figure 4 presents an example LTS network (Fig. 4a) and its corresponding system LTS (Fig. 4b): How LTSs Π_1 and Π_2 should (not) synchronise
is given by the three laws in $\mathcal{V}$: action a needs to be performed by both LTSs
together, leading to an a transition in the system LTS, b and d of Π_1 and Π_2,
respectively, need to synchronise to an e-transition in the system LTS, and c can
be performed by Π_1 independently.

In this setting, model-to-model transformations can be formalised by means
of *LTS transformation rule systems*, inspired by double pushout graph rewriting [24]. Here, we explain the basics by means of an example. A formal treatment
can be found in [51,60]. Figure 5 shows an example LTS $\mathcal{G}$ on the left (Fig. 5a),
with state 1 the initial state. A transformation rule is a pair of LTSs $\langle \mathcal{L}, \mathcal{R} \rangle$, with
$\mathcal{L}$ and $\mathcal{R}$ having some states in common, called the *glue in-states* and *glue exit-
states*. In the example, the round grey states are glue in-states and the square
grey states are glue exit-states. The example rule (Fig. 5b) expresses the following: a sequence of two a-transitions should be replaced by a τ-transition, followed
in sequence by two a'-transitions. Moreover, the rule can only be matched on a
sequence of two a-transitions in $\mathcal{G}$ if that sequence satisfies the following criteria:

1. The first state does not have any additional outgoing transitions (expressed
   by the glue in-state of $\mathcal{L}$);
2. The last state does not have any additional incoming transitions (expressed
   by the glue exit-state of $\mathcal{L}$);
3. The intermediate state has no additional in- or outgoing transitions
   (expressed by the fact that state $\tilde{2}$ of $\mathcal{L}$ is not present in $\mathcal{R}$, meaning that
   a state matched on $\tilde{2}$ is supposed to be removed and replaced by two new

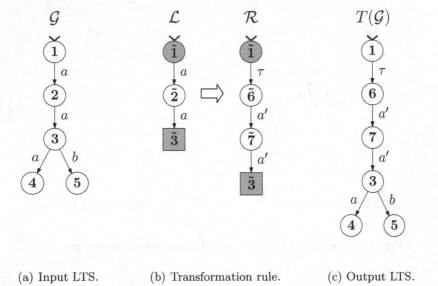

(a) Input LTS.   (b) Transformation rule.   (c) Output LTS.

Fig. 5. An example of applying an LTS transformation rule on an LTS. Round grey state: glue in-state, square grey state: glue exit-state.

states $\tilde{6}$ and $\tilde{7}$, plus the fact that the removal of states in an LTS may not affect its transitions in ways not addressed by the transformation rule).

The rule $\langle \mathcal{L}, \mathcal{R} \rangle$ is applicable on $\mathcal{G}$, and applying it results in the LTS $T(\mathcal{G})$ given in Fig. 5c.

Model-to-model transformations formalised by means of LTS transformation rule systems can be reasoned about without considering a particular LTS on which they are applied. We only assume that such an LTS satisfies particular functional properties of interest. For instance, if an LTS satisfies the property "Always eventually c occurs", expressed in an action-based temporal logic, then it is clear that the transformation rule given in Fig. 5 preserves this property when applied on that LTS, regardless of the latter's structure. The property is preserved as the rule does not transform c-transitions, nor does it affect the reachability of states in the LTS.

The model-to-model transformation verification technique developed in [50, 51, 62] considers functional properties expressed in the modal μ-calculus [39]. Given a μ-calculus formula φ, actions in $\mathcal{L}$ and $\mathcal{R}$ of a transformation rule are automatically abstracted away, i.e., replaced by the silent action τ, if they are considered irrelevant for φ [45]. After this, if $\mathcal{L}$ and $\mathcal{R}$, extended to make explicit that they represent embeddings in a larger LTS on which they are applied, are *divergent-preserving branching bisimilar* [29], which is an equivalence for LTSs sensitive to τ-transitions while still considering the branching structure and τ-loops of the LTSs, then it can be concluded that φ will be satisfied by the LTS produced by the transformation. Although highly non-trivial, this can be extended to LTS transformation rule systems that transform the synchronising behaviour between LTSs. In [51], a formalisation of this technique is presented in detail that has been proven correct with the Coq Proof Assistant [9].

Performance-wise, model-to-model transformation verification is a great improvement over reverifying models. All the considered example transformations could be verified w.r.t. property-preservation practically instantly [51]. The approach has one main drawback, though: sometimes, functional properties can only be expressed, and only become relevant, once the model has obtained a certain amount of detail. If a property must be expressed about the behaviour introduced by one or more model-to-model transformations, property-preservation is not relevant, and verification of the current model seems inevitable.

To possibly even avoid reverification in such cases, more recently, we conducted research on reasoning about the *effect* of an individual LTS transformation rule r on a system property ψ [21], expressed in Action-based Linear Temporal Logic (ALTL) [27,49], when applied on a component satisfying that property. First, a representative LTS $\mathcal{L}_\psi$ for components satisfying an ALTL formula ψ is constructed, by creating the cross-product of a representative LTS $\mathcal{L}$ for all LTSs on which r is applicable with an action-based Büchi automaton $\mathcal{B}_\psi$ encoding ψ [18]. On $\mathcal{L}_\psi$, r is applied, resulting in action-based Büchi automaton $T(\mathcal{L}_\psi)$. After detecting and removing non-accepting cycles of $T(\mathcal{L}_\psi)$, and minimising the resulting Büchi automaton using standard minimisation techniques [23], a characteristic formula for the action-based Büchi automaton is created in the form of a system of μ-calculus equations. Similar to property-preservation checking, this approach works practically instantly. To generalise it to the setting of [51], in future work, rule systems consisting of multiple LTS transformation rules should be considered, and functional properties written in the modal μ-calculus, more expressive than ALTL, should be supported. This approach is promising, but still restricted to updating temporal logic formulae expressed originally for the initial model. If entirely new properties become relevant for a model at a later stage in the MDSE workflow, verification of that model is unavoidable.

Challenges and Directions for Future Work. The developed technique to formally verify model-to-model transformations reasons about the semantics of component-based systems by focussing on LTSs and their transformation, but a suitable formalism to express SLCO-to-SLCO transformations in a way compatible with this is yet to be identified. One direction for future work is to find such a model transformation language. Existing general-purpose transformation languages, such as ATL and XTEND, may be suitable, or a Domain-Specific Transformation Language could be developed that directly involves the SLCO constructs. Extending the technique to reason about the effect of a transformation on a system property will be further investigated as addressed above.

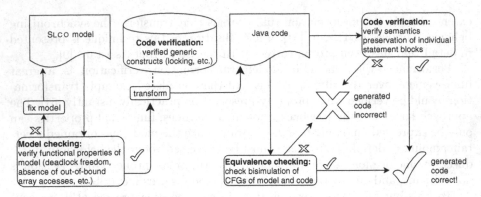

Fig. 6. Verified generation of multi-threaded JAVA code.

4 Verifying Code Generators

When applied on an SLCO model, a code generator should produce code that is (as much as possible) semantically equivalent to the model. In 2014, a new project started, focussed on the model-driven development of multi-threaded software. It was decided to develop a code generator for multi-threaded JAVA code. JAVA being more elaborate than NQC, it was soon clear that SLCO needed to be extended. Version 2.0 of the language [53] introduced the following features:

- Support for the Byte primitive type, and arrays of the primitive types;
- Transition priorities. These allow expressing that the outgoing transitions of a state must be considered for firing in a fixed order;
- An *Action* statement (do <action>): this allows assigning arbitrary action labels to transitions, which can represent particular events, such as calling external functions. Action statements are always enabled.
- A *Composite* statement ([<expr>;<assgn_1>;...;<assgn_n>]): this allows combining certain statements into a single, atomically executed, statement. It starts with an expression, which may be **true**, and a sequence of one or more assignments. The statement is enabled iff the expression is enabled.

The general workflow of producing multi-threaded JAVA code in a verified way is presented in Fig. 6, while an example SLCO model and part of its JAVA implementation is given in Fig. 7. The SLCO-to-JAVA code generator creates one thread for each state machine in the objects of the given SLCO model. In Fig. 7, some of the code for a thread executing the transitions of state machine SM1 of object p is given. Each thread has access to a *lock keeper* (lines 6–7 of the JAVA code), which manages the locks used to avoid data races when accessing object-local variables. This lock keeper uses an ordered locking scheme that prevents circular lock dependencies between threads.

A thread executes according to the associated state machine as follows: the initial state is defined by the constructor method at line 10. In the **exec** method (line 26), the current state of the state machine is repeatedly checked (lines

```
1   model ex_shr {
2     actions a
3     classes
4       P {
5         variables
6           Integer x
7         state machines
8           SM1 {
9             initial R0 states R1
10            transitions
11              R0 -> R1 { [x = 0; x := 1] }
12              R1 -> R0 { do a }
13          }
14          SM2 {
15            initial S0
16            transitions
17              S0 -> S0 { x % 2 = 1 }
18          }
19        }
20      objects p: P()
21  }
```

```
1   ...
2   class java_SM1Thread extends Thread {
3     private Thread java_t;
4     // Current state
5     private ex_shr.java_State java_cState;
6     // Keeper of global variables
7     private ex_shr.java_Keeper java_kp;
8
9     // Constructor
10    java_SM1Thread (ex_shr.java_Keeper java_k) {
11      java_cState = ex_shr.java_State.R0;
12      java_kp = java_k;
13    }
14
15    // Transition functions
16    boolean execute_R0_0() {
17      // [ x = 0; x := 1 ] ... Acquire locks ...
18      if (!(x == 0)) { java_kp.unlock(1); return false; }
19      x = 1;
20      java_kp.unlock(1);
21      return true;
22    }
23    boolean execute_R1_0() { a(); return true; }
24
25    // Execute method
26    public void exec() {
27      while(true) {
28        switch(java_cState) {
29          case ex_shr.java_State.R0:
30            if (execute_R0_0()) { java_cState = ex_shr.java_State.R1; }
31            break;
32          case ex_shr.java_State.R1:
33            if (execute_R1_0()) { java_cState = ex_shr.java_State.R0; }
34            break;
35          default: return;
36  }}}
37  ...
```

Fig. 7. An SLCO shared memory system (top) and derived JAVA code (bottom).

27–28), and depending on its value, one or more functions are executed that correspond one-to-one with a transition in the SLCO model. At lines 16–22, the function `execute_R0_0` is given, which corresponds with the transition at line 11 of the SLCO model. First, it is attempted to acquire a lock for variable x. Once this is achieved, it is checked whether the expression of the composite statement evaluates to **true** (line 18). If it does not, the lock is released and **false** is returned. If it does, x is updated, the lock released, and **true** is returned. If a transition function returns **true**, the thread updates its state and continues checking the current state. Note at line 23 that the action a is mapped to some external function with the same name.

Complete formal verification of a code generator is very challenging [72]. First, we focussed on proving correctness of the *model-independent* parts of the code: we proved that the lock keeper does not introduce deadlocks due to threads waiting for each other [73], that the JAVA channels work as specified by the SLCO channels [16], and that a safety construct called *Failbox* works as intended [15]. For this, the VERIFAST code verifier was used, which allows verifying that JAVA code adheres to pre- and post-conditions specified in separation logic [34]. These verified constructs can be safely used in generated code (see Fig. 6).

The next step was to verify *model-specific* code. Verifying that a code generator always produces correct model-specific code would require reasoning about all possible inputs, i.e., SLCO models. As this is very complex, we focussed on trying to verify automatically that for a given SLCO model, the produced JAVA code correctly implements it, i.e., adheres to the semantics of the model. We achieved this in a two-step approach [67]: first, the control flows of both a thread and its corresponding state machine are extracted. After some straightforward transformations that bring the two control flow graphs conceptually closer together, they are stored in a common graph structure. It is then checked whether those graphs are *bisimilar*. If they are, then we have established that the thread and the state machine perform their steps in equivalent ways. What remains is to establish that the individual steps of the thread indeed correspond with the individual steps of the state machine. For this, code verification is used again. The individual JAVA transition functions are automatically annotated with pre- and post-conditions in separation logic, expressing the semantics of the corresponding SLCO statements. This time, we used the VERCORS verifier to perform the verification [12]. As the pre- and post-conditions are generated automatically, performing the verification only requires pushing a button.

Finally, we investigated techniques to check whether an implementation would still adhere to SLCO's semantics if a platform with a *weak memory model* was targeted [52]. Such a model allows out-of-order execution of instructions, which may violate the intended functionality. In related work, this problem has been addressed in two different ways: in one, a dependency graph is constructed by statically analysing the code [2,57]. This graph encodes which instructions depend on each other due to them accessing the same variables. Next, cycles in this graph that meet certain criteria, depending on the targeted memory model, represent violations of that model. The other way is to apply model checking, considering both the usual possible executions, with instructions occurring in the specified order, and executions in which the instructions have been reordered, insofar allowed by the memory model [1,44]. The drawback of the first approach is its imprecision, while the drawback of the second approach is a state space explosion that is typically even much worse than in standard model checking.

Our contribution was to combine the two approaches: first, explore the state space of the SLCO model, but only considering the executions with instructions in the specified order, and derive from this a dependency graph. Second, apply cycle detection analysis on this dependency graph. For the state space exploration step, a model-to-model transformation from SLCO to MCRL2 [19] was devised. As the produced graphs tend to be more precise than when using static analysis, the results in our experiments were of higher quality, and the overall runtime was often even faster. As the number of elementary cycles in a graph can grow exponentially, constructing a more accurate dependency graph can avoid introducing many cycles that an over-approximation of the potential behaviour would introduce. This reduced number of cycles greatly impacts the processing time, often compensating for the time it takes to explore the state space.

Challenges and Directions for Future Work. The main challenge in this research line is to achieve full verification of code generators. In related work on compiler and code generator verification, full verification has been achieved with theorem proving [11,14,40,43,58], but this is a labour-intensive approach that is not very flexible w.r.t. updates of the compiler or generator. We plan to work on techniques that allow flexible maintenance of correctness proofs.

Another direction currently investigated involves the generation of code for graphics processing units (GPUs) [30]. For many-core programs, however, SLCO is not directly suitable. *Array languages*, on the other hand, have been designed with parallel array processing in mind, which aligns very well with typical GPU functions. We are currently investigating how to embed program verification into the HALIDE language, a language to express image and tensor computations [54]. This language separates *what* a program should do, i.e., its functionality, from *how* it should do it, i.e., the scheduling that involves performance optimisations. Besides making development more insightful, this separation also positively affects verifiability. Verification of the functional correctness of a program can be separated from verifying that optimisations applied to it preserve that correctness. In other work, we focus on updating pre- and post-conditions when code is automatically optimised, to allow for push-button reverification of the code [56].

5   GPU-Accelerated Model Checking

We addressed the verification of model-to-model transformations and code generators, but proving their correctness ultimately depends on the input models being correct. Hence, verifying the correctness of SLCO models cannot be avoided, and sometimes needs to be performed multiple times in an MDSE workflow, as discussed in Sect. 3.2. Initially, we developed an SLCO-to-MCRL2 transformation

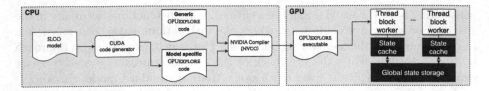

Fig. 8. The workflow from SLCO model to GPUEXPLORE model checking.

for this, to apply the MCRL2 toolset [19] for the model checking of (untimed) SLCO models. Recently, we integrated SLCO in another line of research that started in 2013, focussed on accelerating model checking with GPUs [20,47,69–71].

The research on GPU-acceleration of model checking is motivated first of all by the fact that for a seamless integration of formal verification in MDSE, it is crucial that models can be verified efficiently. If verification takes a long time, this hinders development. Second of all, as hardware developments are increasingly focussed on dedicated devices such as GPUs and adding cores to processors, as opposed to making individual cores faster, computationally intensive computations, such as model checking, require massively parallel algorithms [42].

Figure 8 presents the workflow of formally verifying the correctness of SLCO models with the model checker GPUEXPLORE version 3.0 [65,66]. First, an SLCO model is analysed by a CUDA code generator. CUDA, the Compute Unified Device Architecture, is a parallel computing platform and application programming interface developed by NVIDIA, that can be used to develop programs for their GPUs. The generator produces CUDA C++ code that implements an explicit-state model checker for that specific SLCO model: it includes generated functions that allow the evaluation and firing of SLCO transitions by directly executing instructions corresponding with the associated SLCO statements.

The generated code can be compiled with NVIDIA's NVCC compiler. Note in Fig. 8 that the compiler combines generic, model-independent code, with model-specific code, similar to the SLCO-to-JAVA transformation (Sect. 4).

On the right of Fig. 8, the main concepts of a GPUEXPLORE program are mapped to a GPU architecture: A GPU consists of many streaming multiprocessors (SM) that each have a limited amount of fast, on-chip *shared* memory, and one shared pool of *global* memory. Typically, a GPU program consists of a program, executed by one or more CPU threads, in which GPU functions, called *kernels*, are launched. These kernels are typically executed by many thousands of threads simultaneously. Threads are grouped into *blocks*. A block is executed by an SM, and the threads in a block share a specified amount of shared memory. It is not possible for the threads in one block to access the shared memory of another block. Finally, all blocks share the global memory. In GPUEXPLORE, this memory is used to maintain a large hash table, in which the SLCO model states are stored as they are reached, starting with the initial state.

Table 2. State space exploration speed of SPIN, LTSMIN and GPUEXPLORE, in millions of states per second. -O.M.-: out of memory (32 GB).

Model	Nr. states	SPIN 4-core	LTSMIN 4-core	GPUEXPLORE
adding.50+	529,767,730	-O.M.-	5.36	**148.28**
anderson.6	18,206,917	1.36	1.31	**31.57**
at.6	160,589,600	0.87	2.39	**40.56**
frogs.5	182,772,126	1.05	2.63	**10.31**
lamport.8	62,669,317	1.78	2.19	**34.92**
peterson.6	174,495,861	0.76	2.45	**33.58**
szymanski.5	79,518,740	1.57	1.82	**18.34**

An SLCO model state is a vector defining a state of the model, i.e., it defines for each state machine its current state, and for each variable its current value. Each block repeatedly fetches *unexplored* states, i.e., states for which the outgoing transitions of the corresponding current states of the state machines have not yet been considered for firing. Exploring these states leads to the creation of *successors*, i.e., states reachable by firing a transition. This is conducted in parallel by the threads in a block; GPUEXPLORE runs blocks of 512 threads each. Successors are temporarily stored in shared memory, in which a block-local hash table is maintained. This prevents blocks from frequently accessing slow global memory (which is typically a major performance bottleneck). Once a batch of new successors has been generated, their presence in the global memory hash table is checked. States not yet present are added, ready to be explored in the next round. This procedure is repeated until no more states are generated.

Currently, GPUEXPLORE supports deadlock checking, with support for the verification of Linear Temporal Logic (LTL) [49] formulae being planned for the near future. Table 2 presents some results obtained when comparing the state space exploration speed of GPUEXPLORE with SPIN and the model checker LTSMIN [36]. Both SPIN and LTSMIN support CPU multi-core explicit-state model checking. The models listed here are all SLCO models obtained by translating the model in the BEEM benchmark suite [48] of the same name from the DVE language to SLCO, except for adding.50+, which was obtained by scaling up the adding models present in that benchmark suite. We used a machine with a four-core CPU i7-7700 (3.6 GHz), 32 GB RAM, and an NVIDIA Titan RTX GPU with 24 GB global memory, running LINUX MINT 20 and CUDA 11.4.

As LTSMIN achieves near-linear speedups as the number of used cores is increased [59], these numbers indicate how many cores would be needed to match the speed of GPUEXPLORE. GPUEXPLORE can reach impressive speeds up to 148 million states per second. However, what stands out is that the achieved speed differs greatly between models, more than with SPIN and LTSMIN. In the near future, we will inspect the models and their state spaces, to identify the cause for these differences, and improve the reliability of GPUEXPLORE.

Challenges and Directions for Future Work. The first aspect to address is the verification of temporal logic formulae. First, we will focus on LTL. However, state-of-the-art sequential LTL verification algorithms rely on Depth-First Search (DFS) of the state spaces, as they involve cycle detection. Since DFS is not suitable for GPUs, GPUEXPLORE applies a greedy, Breadth-First Search based exploration algorithm, in which cycle detection cannot be integrated as straightforwardly. In the past, we have investigated algorithms for this that are incomplete [68]. Designing an alternative that is complete remains a challenge.

Another line of research is to achieve GPU acceleration of *probabilistic* model checking [8]. In the past, this has been partially accelerated with GPUs [13,17,37,63]: once the state space has been generated, verification of a probabilistic property, formalised in Probabilistic Computation Tree Logic [31], involves repeated matrix-vector multiplications, which GPUs can perform very rapidly. Also accelerating the state space generation will likely be a major step forward, not only because the generation itself will become faster, but also because it will remove the need to transfer a matrix, representing the state space, from the main memory to the GPU memory.

6   Conclusions

We presented an overview of the research conducted in the last decade on integrating formal verification into an MDSE workflow centered around the SLCO DSL. For an effective integration, efficient verification of models, model-to-model transformations and code generators is crucial. In the three research lines focussing on each of these three types of MDSE artefacts, important steps have been made, and open challenges remain for the (near) future.

One particular challenge bridging the first two lines concerns identifying ways to combine model verification and model-to-model transformation verification, ideally to achieve an automatic verification technique that, depending on the property, the model, and the transformation to be applied, can derive how the transformed model relates to that property. We envision that in order to derive this, a number of verification results need to be established, of which some could possibly be determined via model verification, while for others, model-to-model transformation verification could be more efficient.

References

1. Abdulla, P.A., Atig, M.F., Ngo, T.-P.: The best of both worlds: trading efficiency and optimality in fence insertion for TSO. In: Vitek, J. (ed.) ESOP 2015. LNCS, vol. 9032, pp. 308–332. Springer, Heidelberg (2015). https://doi.org/10.1007/978-3-662-46669-8_13
2. Alglave, J., Kroening, D., Nimal, V., Poetzl, D.: Don't sit on the fence: a static analysis approach to automatic fence insertion. ACM Trans. Progr. Lang. Syst. **39**(2), 6 (2017)

3. Amrani, M., et al.: Formal verification techniques for model transformations: a tridimensional classification. J. Object Technol. **14**(3), 1–43 (2015). https://doi.org/10.5381/jot.2015.14.3.a1

4. van Amstel, M.: Assessing and improving the quality of model transformations. Ph.D. thesis, Eindhoven University of Technology (2011)

5. van Amstel, M., van den Brand, M., Engelen, L.: An exercise in iterative domain-specific language design. In: EVOL/IWPSE, pp. 48–57. ACM Press (2010)

6. van Amstel, M., van den Brand, M., Engelen, L.: Using a DSL and fine-grained model transformations to explore the boudaries of model verification. In: MVV, pp. 120–127. IEEE Computer Society Press (2011)

7. van Amstel, M., van den Brand, M., Protić, Z., Verhoeff, T.: Model-driven software engineering. In: Hamberg, R., Verriet, J. (eds.) Automation in Warehouse Development, pp. 45–58. Springer, London (2011). https://doi.org/10.1007/978-0-85729-968-0_4

8. Baier, C., Katoen, J.: Principles of Model Checking. MIT Press, Cambridge (2008)

9. Bertot, Y., Castéran, P.: Interactive Theorem Proving and Program Development, Coq' Art: The Calculus of Inductive Constructions. Texts in Theoretical Computer Science. An EATCS Series, Springer, Heidelberg (2004). https://doi.org/10.1007/978-3-662-07964-5

10. Bettini, L.: Implementing Domain-Specific Languages with Xtext and Xtend, 2nd edn. Packt Publishing, Birmingham (2016)

11. Blech, J., Glesner, S., Leitner, J.: Formal verification of java code generation from UML models. In: Fujaba Days 2005, pp. 49–56 (2005)

12. Blom, S., Darabi, S., Huisman, M., Oortwijn, W.: The VerCors tool set: verification of parallel and concurrent software. In: Polikarpova, N., Schneider, S. (eds.) IFM 2017. LNCS, vol. 10510, pp. 102–110. Springer, Cham (2017). https://doi.org/10.1007/978-3-319-66845-1_7

13. Bošnački, D., Edelkamp, S., Sulewski, D., Wijs, A.: Parallel probabilistic model checking on general purpose graphics processors. STTT **13**(1), 21–35 (2011). https://doi.org/10.1007/s10009-010-0176-4

14. Bourke, T., Brun, L., Dagand, P.E., Leroy, X., Pouzet, M., Rieg, L.: A formally verified compiler for Lustre. In: PLDI. ACM SIGPLAN Notices, vol. 52, pp. 586–601. ACM (2017)

15. Bošnački, D., et al.: Dependency safety for java: implementing failboxes. In: PPPJ: Virtual Machines, Languages, and Tools, pp. 15:1–15:6. ACM (2016)

16. Bošnački, D., et al.: Towards modular verification of threaded concurrent executable code generated from DSL models. In: FACS, pp. 141–160 (2015)

17. Bošnački, D., Edelkamp, S., Sulewski, D., Wijs, A.: GPU-PRISM: an extension of PRISM for general purpose graphics processing units. In: PDMC, pp. 17–19. IEEE (2010). https://doi.org/10.1109/PDMC-HiBi.2010.11

18. Büchi, J.: On a decision method in restricted second order arithmetic. In: CLMPS, pp. 425–435. Stanford University Press (1962)

19. Bunte, O., et al.: The mCRL2 toolset for analysing concurrent systems. In: Vojnar, T., Zhang, L. (eds.) TACAS 2019. LNCS, vol. 11428, pp. 21–39. Springer, Cham (2019). https://doi.org/10.1007/978-3-030-17465-1_2

20. Cassee, N., Neele, T., Wijs, A.: On the scalability of the GPUexplore explicit-state model checker. In: GaM. EPTCS, vol. 263, pp. 38–52. Open Publishing Association (2017)

21. Chaki, R., Wijs, A.: Formally characterizing the effect of model transformations on system properties. In: Tapia Tarifa, S.L., Proença, J. (eds.) FACS 2022. LNCS,

vol. 13712, pp. 39–58. Springer, Cham (2022). https://doi.org/10.1007/978-3-031-20872-0_3

22. Di Ruscio, D., Kolovos, D., de Lara, J., Pierantonio, A., Tisi, M., Wimmer, M.: Low-code development and model-driven engineering: two sides of the same coin? Softw. Syst. Model. **21**, 437–446 (2022)

23. Duret-Lutz, A., Lewkowicz, A., Fauchille, A., Michaud, T., Renault, É., Xu, L.: Spot 2.0 — a framework for LTL and ω-automata manipulation. In: Artho, C., Legay, A., Peled, D. (eds.) ATVA 2016. LNCS, vol. 9938, pp. 122–129. Springer, Cham (2016). https://doi.org/10.1007/978-3-319-46520-3_8

24. Ehrig, H., Pfender, M., Schneider, H.: Graph-grammars: an algebraic approach. In: SWAT, pp. 167–180. IEEE Computer Society Press (1973)

25. Engelen, L.: From napkin sketches to reliable software. Ph.D. thesis, Eindhoven University of Technology (2012)

26. Feijen, W., van Gasteren, A.: The alternating bit protocol. In: Feijen, W., van Gasteren, A. (eds.) On a Method of Multiprogramming. Monographs in Computer Science, pp. 333–345. Springer, New York (1999). https://doi.org/10.1007/978-1-4757-3126-2_30

27. Giannakopoulou, D.: Model checking for concurrent software architectures. Ph.D. thesis, University of London (1999)

28. Giese, H., Lambers, L.: Towards automatic verification of behavior preservation for model transformation via invariant checking. In: Ehrig, H., Engels, G., Kreowski, H.-J., Rozenberg, G. (eds.) ICGT 2012. LNCS, vol. 7562, pp. 249–263. Springer, Heidelberg (2012). https://doi.org/10.1007/978-3-642-33654-6_17

29. van Glabbeek, R., Luttik, S., Trčka, N.: Branching bisimilarity with explicit divergence. Fundam. Inf. **93**(4), 371–392 (2009)

30. van den Haak, L.B., Wijs, A., van den Brand, M., Huisman, M.: Formal methods for GPGPU programming: is the demand met? In: Dongol, B., Troubitsyna, E. (eds.) IFM 2020. LNCS, vol. 12546, pp. 160–177. Springer, Cham (2020). https://doi.org/10.1007/978-3-030-63461-2_9

31. Hansson, H., Jonsson, B.: A logic for reasoning about time and reliability. Formal Aspects Comput. **6**(5), 512–535 (1994)

32. Holzmann, G.: The model checker spin. IEEE Trans. Software Eng. **23**(5), 279–295 (1997). https://doi.org/10.1109/32.588521

33. Hülsbusch, M., König, B., Rensink, A., Semenyak, M., Soltenborn, C., Wehrheim, H.: Showing full semantics preservation in model transformation - a comparison of techniques. In: Méry, D., Merz, S. (eds.) IFM 2010. LNCS, vol. 6396, pp. 183–198. Springer, Heidelberg (2010). https://doi.org/10.1007/978-3-642-16265-7_14

34. Jacobs, B., Smans, J., Philippaerts, P., Vogels, F., Penninckx, W., Piessens, F.: VeriFast: a powerful, sound, predictable, fast verifier for C and Java. In: Bobaru, M., Havelund, K., Holzmann, G.J., Joshi, R. (eds.) NFM 2011. LNCS, vol. 6617, pp. 41–55. Springer, Heidelberg (2011). https://doi.org/10.1007/978-3-642-20398-5_4

35. Jouault, F., Kurtev, I.: Transforming models with ATL. In: Bruel, J.-M. (ed.) MODELS 2005. LNCS, vol. 3844, pp. 128–138. Springer, Heidelberg (2006). https://doi.org/10.1007/11663430_14

36. Kant, G., Laarman, A., Meijer, J., van de Pol, J., Blom, S., van Dijk, T.: LTSmin: high-performance language-independent model checking. In: Baier, C., Tinelli, C. (eds.) TACAS 2015. LNCS, vol. 9035, pp. 692–707. Springer, Heidelberg (2015). https://doi.org/10.1007/978-3-662-46681-0_61

37. Khan, M.H., Hassan, O., Khan, S.: Accelerating SpMV multiplication in probabilistic model checkers using GPUs. In: Cerone, A., Ölveczky, P.C. (eds.) ICTAC 2021. LNCS, vol. 12819, pp. 86–104. Springer, Cham (2021). https://doi.org/10.1007/978-3-030-85315-0_6

38. Kleppe, A., Warmer, J., Bast, W.: MDA Explained: The Model Driven Architecture(TM): Practice and Promise. Addison-Wesley Professional, Boston (2005)

39. Kozen, D.: Results on the propositional μ-calculus. Theor. Comput. Sci. **27**(3), 333–354 (1983)

40. Kumar, R., Myreen, M., Norrish, M., Owens, S.: CakeML: a verified implementation of ML. In: POPL. ACM SIGPLAN Notices, vol. 49, pp. 179–191. ACM (2014)

41. Lang, F.: Exp.Open 2.0: a flexible tool integrating partial order, compositional, and on-the-fly verification methods. In: Romijn, J., Smith, G., van de Pol, J. (eds.) IFM 2005. LNCS, vol. 3771, pp. 70–88. Springer, Heidelberg (2005). https://doi.org/10.1007/11589976_6

42. Leiserson, C.E., et al.: There's plenty of room at the top: what will drive computer performance after Moore's law? Science **368**(6495) (2020). https://doi.org/10.1126/science.aam9744

43. Leroy, X.: Formal proofs of code generation and verification tools. In: Giannakopoulou, D., Salaün, G. (eds.) SEFM 2014. LNCS, vol. 8702, pp. 1–4. Springer, Cham (2014). https://doi.org/10.1007/978-3-319-10431-7_1

44. Linden, A., Wolper, P.: A verification-based approach to memory fence insertion in PSO memory systems. In: Piterman, N., Smolka, S.A. (eds.) TACAS 2013. LNCS, vol. 7795, pp. 339–353. Springer, Heidelberg (2013). https://doi.org/10.1007/978-3-642-36742-7_24

45. Mateescu, R., Wijs, A.: Property-dependent reductions adequate with divergence-sensitive branching bisimilarity. Sci. Comput. Program. **96**(3), 354–376 (2014)

46. Narayanan, A., Karsai, G.: Towards verifying model transformations. In: Proceedings of 7th International Workshop on Graph Transformation and Visual Modeling Techniques (GT-VMT 2008). ENTCS, vol. 211, pp. 191–200. Elsevier (2008)

47. Neele, T., Wijs, A., Bošnački, D., van de Pol, J.: Partial-order reduction for GPU model checking. In: Artho, C., Legay, A., Peled, D. (eds.) ATVA 2016. LNCS, vol. 9938, pp. 357–374. Springer, Cham (2016). https://doi.org/10.1007/978-3-319-46520-3_23

48. Pelánek, R.: BEEM: benchmarks for explicit model checkers. In: Bošnački, D., Edelkamp, S. (eds.) SPIN 2007. LNCS, vol. 4595, pp. 263–267. Springer, Heidelberg (2007). https://doi.org/10.1007/978-3-540-73370-6_17

49. Pnueli, A.: The temporal logic of programs. In: 18th Annual Symposium on Foundations of Computer Science (FOCS), pp. 46–57. IEEE Computer Society (1977)

50. de Putter, S., Wijs, A.: Verifying a verifier: on the formal correctness of an LTS transformation verification technique. In: Stevens, P., Wąsowski, A. (eds.) FASE 2016. LNCS, vol. 9633, pp. 383–400. Springer, Heidelberg (2016). https://doi.org/10.1007/978-3-662-49665-7_23

51. de Putter, S., Wijs, A.: A formal verification technique for behavioural model-to-model transformations. Form. Asp. Comput. **30**(1), 3–43 (2018). https://link.springer.com/article/10.1007/s00165-017-0437-z

52. de Putter, S., Wijs, A.: Lock and fence when needed: state space exploration + static analysis = improved fence and lock insertion. In: Dongol, B., Troubitsyna, E. (eds.) IFM 2020. LNCS, vol. 12546, pp. 297–317. Springer, Cham (2020). https://doi.org/10.1007/978-3-030-63461-2_16

53. de Putter, S., Wijs, A., Zhang, D.: The SLCO framework for verified, model-driven construction of component software. In: Bae, K., Ölveczky, P.C. (eds.) FACS 2018. LNCS, vol. 11222, pp. 288–296. Springer, Cham (2018). https://doi.org/10.1007/978-3-030-02146-7_15

54. Ragan-Kelley, J., et al.: Halide: decoupling algorithms from schedules for high-performance image processing. Commun. ACM **61**(1), 106–115 (2017). https://doi.org/10.1145/3150211

55. Rahim, L., Whittle, J.: A survey of approaches for verifying model transformations. Softw. Syst. Model. 1–26 (2013). https://doi.org/10.1007/s10270-013-0358-0

56. Şakar, Ö., Safari, M., Huisman, M., Wijs, A.: ALPINIST: an annotation-aware GPU program optimizer. In: TACAS 2022. LNCS, vol. 13244, pp. 332–352. Springer, Cham (2022). https://doi.org/10.1007/978-3-030-99527-0_18

57. Shasha, D., Snir, M.: Efficient and correct execution of parallel programs that share memory. ACM Trans. Program. Lang. Syst. **10**(2), 282–312 (1988)

58. Stenzel, K., Moebius, N., Reif, W.: Formal verification of QVT transformations for code generation. In: Whittle, J., Clark, T., Kühne, T. (eds.) MODELS 2011. LNCS, vol. 6981, pp. 533–547. Springer, Heidelberg (2011). https://doi.org/10.1007/978-3-642-24485-8_39

59. van der Vegt, S., Laarman, A.: A parallel compact hash table. In: Kotásek, Z., Bouda, J., Černá, I., Sekanina, L., Vojnar, T., Antoš, D. (eds.) MEMICS 2011. LNCS, vol. 7119, pp. 191–204. Springer, Heidelberg (2012). https://doi.org/10.1007/978-3-642-25929-6_18

60. Wijs, A.: Define, verify, refine: correct composition and transformation of concurrent system semantics. In: Fiadeiro, J.L., Liu, Z., Xue, J. (eds.) FACS 2013. LNCS, vol. 8348, pp. 348–368. Springer, Cham (2014). https://doi.org/10.1007/978-3-319-07602-7_21

61. Wijs, A., Engelen, L.: Efficient property preservation checking of model refinements. In: Piterman, N., Smolka, S.A. (eds.) TACAS 2013. LNCS, vol. 7795, pp. 565–579. Springer, Heidelberg (2013). https://doi.org/10.1007/978-3-642-36742-7_41

62. Wijs, A., Engelen, L.: REFINER: towards formal verification of model transformations. In: Badger, J.M., Rozier, K.Y. (eds.) NFM 2014. LNCS, vol. 8430, pp. 258–263. Springer, Cham (2014). https://doi.org/10.1007/978-3-319-06200-6_21

63. Wijs, A.J., Bošnački, D.: Improving GPU sparse matrix-vector multiplication for probabilistic model checking. In: Donaldson, A., Parker, D. (eds.) SPIN 2012. LNCS, vol. 7385, pp. 98–116. Springer, Heidelberg (2012). https://doi.org/10.1007/978-3-642-31759-0_9

64. Wijs, A., Engelen, L.: Incremental formal verification for model refining. In: MoDeVVa, pp. 29–34. ACM Press (2012)

65. Wijs, A., Osama, M.: GPUexplore 3.0: GPU accelerated state space exploration for concurrent systems with data. In: Caltais, G., Schilling, C. (eds.) SPIN 2023. LNCS, vol. 13872, pp. 188–197. Springer, Cham (2023). https://doi.org/10.1007/978-3-031-32157-3_11

66. Wijs, A., Osama, M.: A GPU tree database for many-core explicit state space exploration. In: Sankaranarayanan, S., Sharygina, N. (eds.) TACAS 2023, Part I. LNCS, vol. 13993, pp. 684–703. Springer, Cham (2023). https://doi.org/10.1007/978-3-031-30823-9_35

67. Wijs, A., Wiłkowski, M.: Modular indirect push-button formal verification of multi-threaded code generators. In: Ölveczky, P.C., Salaün, G. (eds.) SEFM 2019. LNCS, vol. 11724, pp. 410–429. Springer, Cham (2019). https://doi.org/10.1007/978-3-030-30446-1_22

68. Wijs, A.: BFS-based model checking of linear-time properties with an application on GPUs. In: Chaudhuri, S., Farzan, A. (eds.) CAV 2016. LNCS, vol. 9780, pp. 472–493. Springer, Cham (2016). https://doi.org/10.1007/978-3-319-41540-6_26

69. Wijs, A., Bošnački, D.: GPUexplore: many-core on-the-fly state space exploration using GPUs. In: Ábrahám, E., Havelund, K. (eds.) TACAS 2014. LNCS, vol. 8413, pp. 233–247. Springer, Heidelberg (2014). https://doi.org/10.1007/978-3-642-54862-8_16

70. Wijs, A., Bošnački, D.: Many-core on-the-fly model checking of safety properties using GPUs. STTT **18**(2), 169–185 (2016). https://doi.org/10.1007/s10009-015-0379-9

71. Wijs, A., Neele, T., Bošnački, D.: GPUexplore 2.0: unleashing GPU explicit-state model checking. In: Fitzgerald, J., Heitmeyer, C., Gnesi, S., Philippou, A. (eds.) FM 2016. LNCS, vol. 9995, pp. 694–701. Springer, Cham (2016). https://doi.org/10.1007/978-3-319-48989-6_42

72. Zhang, D., et al.: Towards verified java code generation from concurrent state machines. In: AMT@MoDELS, pp. 64–69 (2014)

73. Zhang, D., et al.: Verifying atomicity preservation and deadlock freedom of a generic shared variable mechanism used in model-to-code transformations. In: Hammoudi, S., Pires, L.F., Selic, B., Desfray, P. (eds.) MODELSWARD 2016. CCIS, vol. 692, pp. 249–273. Springer, Cham (2017). https://doi.org/10.1007/978-3-319-66302-9_13

Author Index

J. Cámara and S.-S. Jongmans (Eds.): FACS 2023, LNCS 14485, p. 229, 2024.
https://doi.org/10.1007/978-3-031-52183-6